Technology for Wine and Beer Production from *Ipomoea batatas*

Sandeep Kumar Panda

KIIT University
Bhubaneswar, India

T0144125

CRC Press
Taylor & Francis Group
Boca Raton London New York

CRC Press is an imprint of the
Taylor & Francis Group, an **informa** business

A SCIENCE PUBLISHERS BOOK

CRC Press
Taylor & Francis Group
6000 Broken Sound Parkway NW, Suite 300
Boca Raton, FL 33487-2742

First issued in paperback 2020

© 2019 by Taylor & Francis Group, LLC
CRC Press is an imprint of Taylor & Francis Group, an Informa business

No claim to original U.S. Government works

ISBN-13: 978-0-367-17495-8 (hbk)
ISBN-13: 978-0-367-77933-7 (pbk)

Visit the Taylor & Francis Web site at
http://www.taylorandfrancis.com

and the CRC Press Web site at
http://www.crcpress.com

Preface

<><><><><><><><><><><><><><><><><><><><><><><><><><><><><><><><><><><><><><><>

Several technologies have been developed to improve the quality of food and beverages and frequently several amendments are added from time to time for developing new products of consumers' interest. Fermentation technology is one of the most promising food technologies which contribute to the multibillion dollar sector of alcoholic beverages. This book describes the technological aspects of the development of anthocyanin rich wine, herbal sweet potato wine and an anthocyanin rich beer by using purple sweet potato as substrate. Biochemical study reports have been presented to compare the innovative alcoholic beverages with commercial samples. Also the organoleptic characteristics of the beverages have been depicted. Further, the book describes the techno-economic feasibility for industrial scale production considering the innovative technologies for preparation of wine and beer from purple sweet potato. The book also presents a project report on the investment and cost economics and profitability of an integrated winery and brewery with production capacity of 7500 litre beer per day and 2500 litre wine per day.

Technologies always tend towards fine-tuning and innovations provide novelty and betterment to the society. Considering the aforesaid fact, priority has been given to technology, technoeconomic and socioeconomic impacts of developing the innovative alcoholic beverages from purple sweet potato. I would like to thank the authorities of Regional Centre of Central Tuber Crops Research Institute for their technical support. I am especially thankful to Dr. Shirsendu Banerjee, Assistant Professor, KIIT University and Mr. Bijay Kumar Bej, Chattered Accountant, Ashok Leyland for their support during the preparation of the technoeconomic and profitability report. I hope the readers will like the innovation and the findings of the research. Wish you a happy reading.

Dr. Sandeep K. Panda
School of Biotechnology, KIIT University, India

Contents

List of Abbreviations

ANOVA	:	Analysis of Variance
AOAC	:	Association of Official Analytical Chemists
β	:	Beta
BOD	:	Biological Oxygen Demand
CAE	:	Caffeic Acid Equivalent
$Ca(OH)_2$	:	Calcium hydroxide
CLDL	:	Carbamylated Low Density Lipo-protein
CO_2	:	Carbon dioxide
$CuSO_4$	:	Copper sulphate
DPPH	:	2,2-diphenyl-1picryl hydrazyl
DWB	:	Dry Weight Basis
°C	:	Degree centigrade
FAO	:	Food and Agriculture Organization
FTIR	:	Fourier Transform Infrared
FW	:	Fresh Weight
g	:	Gram
hr	:	Hour
HSD	:	Honestly Significant Difference
H_2O_2	:	Hydrogen peroxide
H_2SO_4	:	Hydrogen sulphate
HCl	:	Hydrogen chloride
Kg	:	Kilogram
K cal	:	Kilo Calorie
l	:	Litre
LA	:	Lactic Acid
LAB	:	Lactic Acid Bacteria
LDL	:	Low Density Lipo-protein
µg	:	Microgram

M	:	Molarity
MCT	:	Mercury Cadmium Telluride
$MgSO_4$	:	Magnesium sulphate
Min	:	Minute
ml	:	Millilitre
$MnSO_4$	:	Manganese sulphate
MT	:	Million Tonnes
MYGP	:	Malt extract Yeast extract Glucose Peptone
N	:	Normality
NMR	:	Nuclear Magnetic Resonance
NS	:	Not Significant
Na_2CO_3	:	Sodium carbonate
Na_2HPO_4	:	Sodium hydrogen phosphate
Na_2SO_4	:	Sodium sulphate
NaCl	:	Sodium chloride
$NaHCO_3$	:	Sodium bicarbonate
NaOH	:	Sodium hydroxide
nm	:	Nanometre
O.D.	:	Optical Density
PCA	:	Principal Component Analysis
PDA	:	Potato Dextrose Agar
%	:	Percentage
PSP	:	Purple Sweet Potato
rpm	:	Rotation per minute
SD	:	Standard Deviation
SP	:	Sweet potato
SPSS	:	Statistical Package for Social Sciences
TA	:	Titratable Acidity
TPC	:	Total Phenolic Content
TS	:	Total Sugars
TSS	:	Total Soluble Solids

1

Introduction

◇◇

Sweet potato (*Ipomoea batatas* L.; Family: *Convolvulaceae*) ranks as the seventh most important food crop of the world (Ray and Ward 2006). Annual sweet potato production worldwide varied from 101.81 million metric tons to 112.84 million metric tons during the year 2012 to 2017 (FAO 2017). China leads in sweet potato production, followed by Malawi and India ranks 10th in the global sweet potato production (http://www.fao.org/faostat/en/#rankings/countries_by_ commodity). China alone contributes around 64% of the global sweet potato production during 2017 (Fig. 1.1).

Forty out of the eighty two (82) developing countries of Africa, Asia and Latin America, where sweet potato is grown, considers it as one of the first five most significant food crops produced yearly (Ray and Balagopalan 1997). India is regarded as one amongst the prime producers of sweet potato in the globe with generation of 1.46 million tons of roots during 2017. Sweet potato is generally planted in poorly fertile, marginal lands with inadequate water availability; still it is observed to produce more calories per area per day as compared to other food crops (Horton et al. 1989). Roots of sweet potato enlarge to produce tuberous roots (storage roots) whereas some roots cannot become storage roots and they remain as pencil roots and absorb water and nutrients from the soil. The tuberous roots are the edible part of the sweet potato plant. Henceforth, sweet potato 'storage roots' would be mentioned as 'roots' instead of tuber

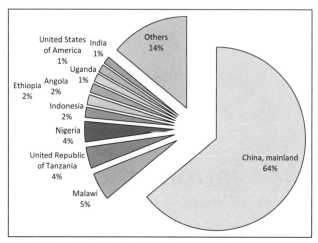

Figure 1.1. Sweet potato production country wise.

(Nedunchezhiyan and Ray 2010) in this book. The fleshy storage roots (Fig. 1.2) are rich in starch content (DW, 50–80% and FW, 7–28%) and the sugar content varies from 4 to15% on dry weight basis (Li et al. 1994). The roots are also enriched with vitamin C, pro-vitamin A, vitamin B and iron (Ray and Tomlins 2010). Presence of colourful

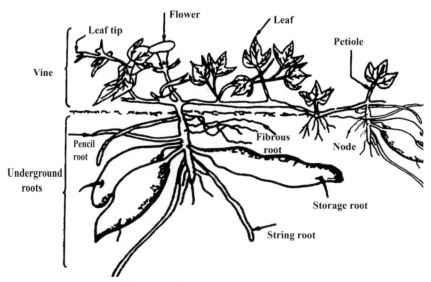

Figure 1.2. Sweet potato storage roots.

2

pigments in the storage roots are the characteristic features of some special varieties of sweet potato. Mostly *β*-carotene and anthocyanin are found to impart colour in these special cultivars (Fig. 1.3) (Woolfe 1992; Ray and Tomlins 2010).

Alcoholic fermentation for wine, beer and other alcoholic beverages is generally carried out by the application of yeast. During the fermentation, glucose is converted into ethanol and carbon dioxide. The entire reaction occurring in the yeast cell is very intricate and the general reaction is presented as:

$$C_6H_{12}O_6 ====> 2(CH_3CH_2OH) + 2(CO_2) + Energy$$
$$\text{(which is stored in ATP)}$$

The chemical representation shows that one mole of glucose translates to two moles of ethyl alcohol and carbon dioxide each, though actually the process is more complex than it appears. Apart from ethyl alcohol and carbon dioxide as principal products, the glucose is biochemically transformed to several byproducts like dead yeast cell mass, glycerol and organic acids like pyruvic acid, ketoglutaric acid, lactic acid, etc. The body of wine is imparted by glycerol. In order to prepare fuller bodied wine and beer, fermentation specialists in wineries and breweries optimize certain physiochemical conditions to generate higher concentration of glycerol as byproduct in the fermentation medium. The sugars used by the yeasts are only the reducing sugars

Figure 1.3. Anthocyanin rich sweet potato variety

Color version at the end of the book

3

such as glucose, fructose, sucrose and maltose. So the sugars in other complex forms should be first converted to the reducing sugars before allowing fermentation.

Natural anthocyanins are known for their protective properties in oxidative stress, microbial infection, non-communicable diseases and many others (Khoo et al. 2017). Several researchers have reported about the high content of anthocyanins with peonidin and cyanidin in purple sweet potatoes. The anthocyanin content differs significantly among various purple sweet potato varieties. A study conducted by Teow et al. (2007) observed a variation in anthocyanin contents (17–531/kg roots) in 19 genotypes. Similarly, another researcher found the anthocyanin content of one cultivar of purple sweet potato to be as high as 1820 mg/kg roots in fresh weight basis (Cevallos-Casals and Cisneros-Zevallos (2003). In a comparative study conducted by Rodriguez-Saona and Wrolstad (2001) it was made out that the anthocyanin content in purple potatoes (*Solanum tuberosum* L.) ranged between 20 to 400 (mg/kg fw) for thirty three varieties. Similarly, the anthocyanin imparted in coloured fruits and vegetables were found to have anthocyanin content in a range of 20–6000 mg/kg fw (Wrolstad 2000). The health promoting potential of purple sweet potato anthocyanins has been depicted in several studies. Saigusa et al. (2005) have demonstrated about the free radical scavenging property of purple sweet potato anthocyanins. Likewise, attenuation of hepatic dysfunction (Han et al. 2007), enhancement of memory (Wu et al. 2008), fall of high glucose level in blood and insulin resistance (Ray and Tomlins 2010) and anti-proliferative attribute against malignant cells (Wang et al. 2006) were the most important and health promoting features of purple sweet potato anthocyanin. Furthermore, anthocyanin pigments of purple sweet potato are of different types, and the specialty exists with the high content of acylated anthocyanins in comparison with its counterparts. The purple sweet potato anthocyanin has a potential to be used in food and beverage processing as the anthocyanin of purple sweet potato is comparatively stable in high temperature and light (Hayashi et al. 2003; Hayashi et al. 1996). Keeping the above health beneficial attributes of purple sweet potato anthocyanin in mind and further studies on animal models by different researchers have proved it as a potential nutraceutical (Khoo et al. 2017). Lim et al. (2013) demonstrated the protective action of anthocyanin rich sweet potato against colorectal cancer in both *in-vitro* and *in-vivo* model as it ceases cell cycle due to anti-proliferative property and apoptosis. Anthocyanin rich sweet potato has been bio-processed to several value added products like curd, lacto-pickle and lacto-juice (Mohapatra et al. 2007; Panda et al. 2009a,b). There are evidences of fermentation of purple sweet potato (rich in anthocyanin) to various alcoholic beverages like wine and beer traditionally in certain areas of Japan and China (Nelson and Elevitch 2011; Odebode et al. 2008). Sweet potato, being rich in starch, vitamins, minerals and anti-oxidants like phenolics, β-carotene and anthocyanin (in some varieties), is highly potential for alcoholic fermentation into wine and beer.

This book deals with:

- Review of production and application of fermented sweet potato food products, fermented biocommodities and unfermented sweet potato products, especially purple sweet potato, throughout the globe.

- Alcoholic fermentation of purple sweet potato into anthocyanin rich wine (red wine) by using *Saccharomyces cerevisiae* as the starter culture.

- Alcoholic fermentation of purple sweet potato along with medicinal and herbal plant parts into herbal wine using *S. cerevisiae* as the starter culture.

- Preparation of beer by using purple sweet potato flakes in different concentrations along with grist (ground malt prepared from barley grains) and hops.

- Biochemical, multivariate statistical and consumer sensory evaluation of purple sweet potato fermented products in the laboratory.

- Technoeconomical feasibility study of anthocyanin rich wine and anthocyanin rich beer developed from purple sweet potato.

Review of Sweet Potato and Its Fermented Products

◇◇

2.1 Root and Tuber Crops in Food Perspective

Roots and tuber crops include cassava (*Manihot esculenta* C.), sweet potato (*Ipomoea batatas* L.), elephant foot yam (*Amorphophallus paeoniifolius*), potato (*Solanum tuberosum*), aroids and other minor root crops such as African yam bean (*Sphenostylis stenocarpa*) and East Indian arrowroot (*Tacca leontopetaloides*). Such crops are reported to serve as an important part of the diet of 2.2 billion people throughout the globe (https://www.nri.org/development-programmes/root-and-tuber-crops-in-development/overview). Root and tuber crops are the cheapest source of dietary energy in terms of carbohydrates in developed countries. Most tuber crops are rich in starch (Moorthy 2004), for example the starch content of sweet potato varies from 12–30% depending upon the variety and the environmental and cultivation methods. Potatoes and yam contain a high amount of protein among other root crops. However, methionine and cystine limit amino acids in tuber crop protein. Several researchers have depicted the nutritional and health beneficial properties of root and tuber crops for example as antioxidative, hypoglycemic, hypocholesterolemic, antimicrobial and immunodialatory activities (Chandrasekara and Kumar 2016). Further, the aforesaid health beneficial potentials of tuber crops may be attributed to several bioactive molecules present in root crops, such as phenolics, saponins and phytic acid. Cassava, sweet potato and yam contain ascorbic acid and some varieties of sweet potato are rich in pigments such as β-carotene and anthocyanin, which impart colour as well as a functional property to the root crop. Innumerable recipes using tuber crops are formulated across the globe. These formulations maybe carried out through cooking or fermentation and are associated with the culture and traditions of a particular region. The popular fermented food products prepared from cassava are *gari, fufu, lafun,* etc. (Panda and Ray 2016). Similarly *shochu* is a traditional alcoholic beverage of Japan prepared from sweet potato, described in the Section 2.4.4. Other fermented foods from sweet potato include lactic acid fermented juice, cubes and yoghurt. Acidophilus milk and vinegar are also successfully prepared by using sweet potato as substrate (please see Section 2.4.2.1.4 and Section 2.4.2.2.1).

2.1.1 Sweet potato

Sweet potato is a tropical tuber crop belonging to the family *Convolvulaceae*. The sweet potato plant is an herbaceous perennial vine having alternate round, triangular or cordate leaves and mid-sized sympetalous flowers. Sweet potato is the 7th most significant crop cultivated on earth and among the root and tuber crops; it follows cassava (Ray and Ravi 2005). Sweet potato was initially illustrated by Linnaeus as *Convolvulus batatas* in 1753. But in 1791 Lanmark classified it in the genus of *Ipomoea* on the criteria of stigma shape and the surface of pollen grains. For this reason, later, it was named *Ipomoea batatas* (L.). Vegetative propagated sweet potato generates adventitious roots that develop into primary fibrous roots that further get divided to lateral roots. Water and necessary nutrients are absorbed by the fibrous roots, which also afford anchorage to the plant and help in the storage of photosynthetic products. With the maturation of the plant, lateral roots develop into storage roots by accumulating photosynthates. Plants grown from the true seeds form a typical tap root in the company of lateral roots. Afterwards, the vital tap root acts as storage root. The stem of sweet potato is cylindrical in shape and the length varies from 1–5 m depending on the cultivar and the availability of water and nutrients. The internode length of the stem differs significantly from short to very long. The leaves are simple and spirally organised on the stem in 2/5 phyllotaxis. The general shape of the leaf may be round, triangular or heart shaped. Sweet potato under normal conditions does not flower. However, some cultivars produce few to large number of flowers (Huaman 1992). The inflorescence is generally a cyme and the flower is bisexual. Calyx consists of five sepals in two whorls (2+3). Corolla comprises of five petals those are infused to form a funnel. The fruit is a capsule and turns brown when matured and can be pubescent or glabrous. Each capsule contains 1–4 seeds of approximately 3 mm in size. The storage roots are of great importance as those develop from adventitious roots. The quantity of latex accumulated is dependent on several factors such as maturity of the storage root, moisture content and the weight gained during the growth period. Storage roots may be formed in clusters or in a dispersed manner depending on the cultivar and they also vary in shape and size. The skin and flesh too vary in colour from white to orange, red and purple (Huaman 1992). Sweet potato grows best and yields storage roots higher a moderately warm climate and temperature of 21–26°C. It shows a tolerance to moderate water scarcity but barely withstand water stress. It requires plenty of sunshine. Yield of storage roots is higher in the dry season as compared to the yield in the rainy season (Nedunchezhiyan and Byju 2005). The ideal soil pH for cultivation of sweet potato is 5.5–6.5. Sweet potato has a great potential of providing energy, proteins, vitamins and minerals and combating food shortages. The mean energy output: input ratios for rice was 17:1 and for sweet potato was 60:1 in Fijian farms (Norman et al. 1984). The major content of the storage roots is moisture which accounts for 70% (average) of the total weight. The compositions of the dry matter of the sweet potato are mainly starch (70%), total sugar (10%), protein (5%), ash (3%), lipids (1%), fibre (10%), vitamins and minerals (<1%) (Woolfe 1992). Among a lot of varieties of sweet potato, the orange fleshed and the purple fleshed sweet potato have a lot of industrial importance. The orange fleshed sweet potato is rich in β-carotene, a precursor of vitamin–A and the purple fleshed sweet potato is rich in anthocyanin.

2.1.1.1 Anthocyanin rich sweet potato

This is a specific and a new variety of sweet potato which is rich in anthocyanin contents. The colour of the flesh of the sweet potato varies from blue to purple and red. Anthocyanin pigments are purely herbal, soluble food colourants that impart red, purple or blue colour to plant or plant parts including fruits, roots and flowers. Anthocyanin has a lot of importance both commercially and from an industrial point of view as it has a broad range of applications due to its brilliant colour, safe, non-toxic, excellent functional attributes and pharmaceutical functions (Rice–Evans and Packer 1998). Anthocyanin is widely used in industries such as pharmaceutical, cosmetics and food colourant. The anthocyanins of sweet potato are comparably better than those present in other vegetables and fruits such as red cabbage. Terahara et al. (2004) have revealed about the pharmaceutical properties of sweet potato anthocyanins such as their free radical scavenging activity, antimutagenic property and its potential to act against hepatic disorders which indicate that purple sweet potato could support maintaining healthy lives of humans. The anthocyanins present in the sweet potato are acylated anthocyanins. The major anthocyanins present in these kinds of sweet potato are cyanidin and peonidin (Lebot 2008). Anthocyanins in sweet potato also play an additional role in restoring the liver functions and blood pressure levels in volunteers with impaired hepatic function or hypertension. The anti-proliferative activities of sweet potato were studied using human lymphoma ND4 cells and it was observed that the water extract of the leaf veins and the storage roots showed the highest anti-oxidative activity. This variety is currently suggested as a better resource for the manufacturing of foods and beverages for human wellness (Suda et al. 2003).

2.2 Origin and Geographical Distribution of Sweet Potato

Sweet potato is a new world crop and is considered to originate from either the central or South American low–lands (Woolfe 1992). So far the oldest dried roots were discovered in the caves of Chilca Canyon of Peru (Engel 1970). Radio–carbonation studies reveal, the roots are of 8000–10000 years old. Similarly Ugent et al. (1982) reported archeological detection of relics of farmed sweet potato from the Casma Valley of Peru, which date back to around 2000 BC. The spreading of sweet potato in historic times was by two lines of transmission: (a) The *Batatas* line which followed after Spain introduced to Europe the sweet potato and continuing till after 1500 AD and through the relocation of European grown clones to African, Indian and the East Indian lands through the Portuguese and (b) Throuh the *Kamote* line where by the Mexican varieties were transferred to Philippines by the Spanish traders (Woolfe 1992). The common names of the sweet potatoes used across the globe are *batatas, tata, mbatata, bombe, bambai and bambaina*. The last three names are linked to the Indian city Bombay (Mumbai) acquired by the British in 1662 which may be linked to a later spread of the plant by the British colonial influences (Woolfe 1992). Sweet potato is distributed and cultivated in nearly all the continents, mostly in Asia and Africa. The distribution of sweet potato in global, Asian and African continents is depicted in Figs. 2.1–2.3.

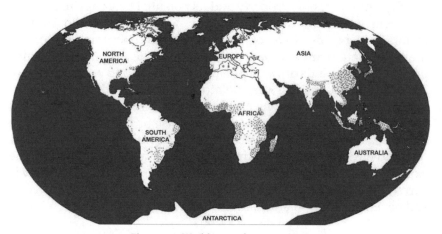

Figure 2.1. World map of sweet potato.

2.3 Overview of Sweet Potato Global Trade

The global sweet potato production varied from 101.81 million metric tons to 105.19 million metric tons during the year 2010–2016. The average production is 103.81 million metric tons during these seven years (https://www.statista.com/statistics/812343/global-sweet-potato-production/). Sweet potato is used differently across different countries and continents. For example sweet potato is used as both food and feed in China. Roasted sweet potato is a popular street food in China. Nigerians and Tanzanians use it as food in the form of a porridge, fermented drinks, crisps, steamed and mashed sweet

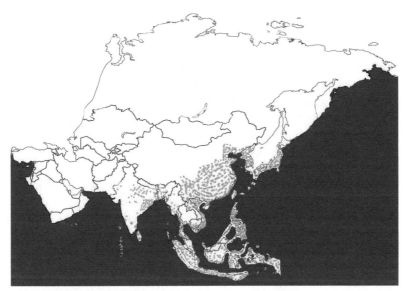

Figure 2.2. Asia map of sweet potato.

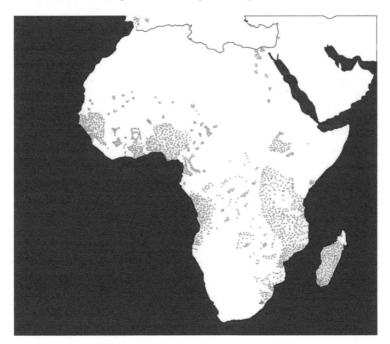

Figure 2.3. Africa map of sweet potato.

potato. Indians use it as a vegetable in curries whereas in Vietnam the preference is different, as the Vietnamese use sweet potato with their meat dishes (https://www.worldatlas.com/articles/world-leaders-in-sweet-potato-production.html). Although sweet potato is cultivated and consumed throughout the world, the trade itself is limited. Europe is a huge importer of sweet potato and mainly the United States is the source. However, European countries like Spain, Portugal, Italy and Greece are known to produce sweet potatoes. UK and the Netherlands are the largest consumers of imported sweet potato and the Netherlands is the transit point for sweet potato for rest of the Europe. The export of sweet potato by the United States is on the rise. In 2015 the export value of sweet potato was 139 million dollars as compared to 66.19 million dollars in 2010. In Spain a 25–30% rise in the cultivation area of sweet potato was observed. In Belgium, the demand is at its highest during the winter and their prime suppliers are the US and Senegal (Mulderji 2016). Egypt exports sweet potato to European countries, mainly to UK. The other importers of Egyptian sweet potato are the Middle East, namely Saudi Arabia, Qatar, Kuwait and the Maldives. The current price is 0.50 Euro per kilo FOB. Belgium offers a good price (1 Euro/kg) to sweet potato imported from Spain, Portugal and US. Australia has an annual production of sweet potato worth 80 million dollars. A small portion of the produce is exported to the UAE and the rest are consumed domestically (Mulderij 2017). The rise of the sweet potato market in US and Europe is attributed to its health promoting constituents such as a high content of vitamin A, vitamin C, fibers and low glycemic index (Bond 2017). After NAFTA (North American Free Trade Agreement) came to force in 1994 the trade volume of sweet potato has increased by more than 500%. However, the price of sweet potato is still significantly different in USA, Canada and Mexico (Lee and Kennedy 2016). In South Africa, Northern Cape, Western Cape, Limpopo, etc., are the provinces cultivating sweet potato. However, South Africa is not considered as a prominent exporter of sweet potato, it has only a stake of 0.4% of the global sweet potato export. The sweet potato produced in South Africa is mostly consumed domestically however: a small quantity of the total produce reaches the Netherlands, UK, Namibia, Botswana (South African Sweet Potato Market Value Chain, 2017).

2.4 Fermentation of Sweet Potato

Fermentation is regarded as the earliest and one of the most prominent food processing methods. It has several nutritional, ethical and health benefits such as removal of anti-nutritional components, overcoming the problems of perishability, increasing the shelf–life of the food products and adding value for enhancing nutrients, taste and flavour as well as generating employment. Indigenous fruit–based fermented products such as wine beer like beverages are prepared and drunk by various communities of the world hundreds of years ago and each type of drink are closely related to the culture of the respective community and region (Battcock and Azam–Ali 2001).

However, it is quite important to gather scientific understanding on the fermentation of fruits and vegetables as the traditional processes which don't always run in a uniform manner, hence the yield, quality and safety of the final product is at stake.

The major challenges in the fermentation of traditional fermented food products that need to be addressed are:

- Improving the understanding of the traditional fermented fruit products, and
- Refining the process

Several fermented beverages such as wine, beer and other alcoholic beverages developed across the globe are currently understood in terms of microbial diversity and biochemical constituents. It has further been observed that the substrates, starter cultures and biochemical conversions vary from one geographical region to other. Hence, the severe variations among alcoholic beverages are observed in Asian, African and Latin American regions. It is not always the objective of food technologists to optimize technology for higher production; rather, it is intended to ensure a high quality of products and safety.

2.4.1 Scope of fermentation for sweet potato

Fermentation is the most well adapted technology for food preservation throughout the world. Fermentation is carried out by microorganisms; which are mainly responsible for the conversion of fruits, vegetables and cereals to various food products like wine, beer, pickles, etc. Industries related to fermentation use specific microorganisms for specific products and a standardised protocol is followed for reaching the desirable attributes of the final product. A quantity of research has been conducted on the fermentation of sweet potato as it contains 70% (dry mass) of starch, 10% (dry mass) of the total sugar and other nutraceutical compounds. The high starch and sugar content have encouraged scientists to study fermentation as microorganisms primarily require a sugar source/substrate.

2.4.2 Bacterial fermentation of sweet potato

Bacteria are a large group of microorganisms. Despite their activity in food spoilage, they have also become a very important tool for fermentation. The beneficial bacteria often used for fermentation of sweet potatoes have been broadly classified into two major groups.

a) Lactic acid bacteria
b) Acetic acid bacteria

2.4.2.1 Fermentation of sweet potato by lactic acid bacteria

Lactic acid bacteria are chosen for the fermentation of sugar substrates such as glucose and fructose to cellular energy and the metabolic by–product lactate/lactic acid. Lactic acid fermentation is the simplest type of fermentation. Lactic acid bacteria play a significant role in the food industry. They utlilise carbon sources from the raw material to produce lactic acid and simultaneously reduce the pH in the fermentation medium. LABS are understood to be generally regarded as safe (GRAS). Fermentation using lactic acid bacteria improves the nutritional value and organoleptic properties of food. This group of bacteria is known to be benevolent towards human as they possess probiotic properties. Some important characteristics of the probiotic property are viability under varied pH conditions, bile salt tolerance, antibiotic resistance and hydrophobic properties. Lactic acid bacteria has applications in both the dairy and non-dairy food sectors. For example, yoghurt is produced using *Streptococcus thermophilus* and *Lactobacillus delbrueckii,* and fermented olives are produced by using *Lb. pentosus* and *Lb. plantarum*. Certain lactic acid bacteria are known to produce antimicrobial compounds to inhibit unwanted microorganisms and impart a particular flavour by producing certain flavouring compounds. *Lb. fermentum* imparts a characteristic flavour during cocoa fermentation and aroma producing compounds such as diacetyl, acetaldehyde and esters are produced by lactic acid bacteria (Kandasamy et al. 2018). By using lactic acid bacteria several food products have been developed from sweet potatoes which have been discussed below.

2.4.2.1.1 Lacto–pickles from sweet potato

Lactobacillus plantarum is the starter culture often applied for the lacto–fermentation of sweet potato, as well as other plant materials (Ray and Panda 2007). Lactic acid bacteria impact the flavour of respective fermented products in various manners. Mostly acids are produced during lactic acid fermentation. Thus the sourness increases in the sweet potatoes and their sweetness decreases as the sugars are fermented to acids. The lacto–pickles have been prepared both from β-carotene as well as anthocyanin rich sweet potato roots.

Pickling was conducted by brining and blanching anthocyanin rich sweet potato cubes, followed by lacto–fermentation using the probiotic strain *plantarum* MTCC 1407. The fermentation process was allowed for 28 days. The lacto-pickle obtained after the fermentation had the following biochemical composition: pH varied from 2.5–2.8, titratable acidity ranges from 1.5–1.7 g/kg, lactic acid content of 1.0–1.3 g/kg, starch, 56–58 g/kg and anthocyanin of 390 mg/kg on pickle (fw). Anthocyanin rich pickle was found acceptable based on organoleptic evaluation (Panda et al. 2009a). Likewise, a β-carotene rich sweet potato pickle has been prepared and sensory analysis showed the acceptability of the product (Panda et al. 2007). The preservative and other additives used with this pickle are soy sauce, sugar, sesame seeds and chillies respectively (Woolfe 1992).

2.4.2.1.2 Lacto juice from sweet potato

Lacto juice has also been prepared by fermenting the juice of both anthocyanin and β-carotene rich sweet potato varieties. Panda and Ray

(2007) and Panda et al. (2009b) have demonstrated the fermentation methodology by using *Lactobacillus plantarum* 1407. There was no significant difference in the biochemical parameters with boiled and non–boiled sweet potato fermented juices except for the β-carotene content. The juices were accepted by the sensory panelists.

2.4.2.1.3 Sweet potato curd and yoghurt

Generally curd and yoghurt are produced by lacto–fermentation of milk and are known as advantageous over milk in nutraceutical and digestive properties (Berger et al. 1979). There are also some instances where milk is fermented along with other adjuncts such as fibres, starch, vegetables such as soybean and sweet potato for the production of better enriched curds and yoghurts. Curd is quite popular in Asian countries while yoghurt is trendy in the USA and the Europe (Sarkar et al. 1996; Younus et al. 2002). Yoghurt like products have been developed by using a sweet potato puree, milk, sucrose and dried inoculums. The product had 0.85% titratable acidity (TA). Rates of TA development decreased with an increase in sweet potato and sugar concentration. The yoghurt was fermented for around 6.5 hr. A trained panelist scored a mean value of 7.7 out of a scale 1–10 (Collins et al. 1991). Likewise, purple sweet potato was used as an ingredient for the preparation of sweet potato curd. Purple sweet potato puree was boiled and blended with milk and further the fermentation was conducted using *Lactobacillus bulgaris*, *Streptococcus lactis* and *Diacetic lactis*. Addition of anthocyanin rich sweet potato puree improved the quality of the curds various attributes such as flavour, texture, mineral contents, nutrients, anti–diabetic components, anthocyanin and dietary fibre. Curd prepared by using 8–12% of sweet potato puree was the most preferred curd by the tasting panel (Panda et al. 2006). In a similar process curd was prepared by using a β-carotene rich sweet potato puree, cow milk and a curd starter. The curd prepared by the addition of 12–16% of β-carotene rich sweet potato puree was the curd most preferred among other combinations. The incorporation of β-carotene rich sweet potato puree between 12 to 16% could make the curd more compact and improve the flavour, β-carotene pigments (antioxidants), etc. (Mohapatra et al. 2007).

2.4.2.1.4 Acidophilus milk from sweet potato

Acidophilus milk is a specialized health promoting beverage which is produced by fermenting milk with *Lactobacillus acidophilus*. Several researchers have carried out in-depth studies on the therapeutic effects of the acidophilus milk and also the interaction of microorganisms in the human intestine. Perez and Tan (2006) have described the potential of the microflora of acidophilus milk in inhibiting pathogenic microorganisms and counteracting against toxin producing harmful microorganisms especially in children (Ray and Panda 2007). Furthermore, purple sweet potato puree has been used in the formulation of acidophilus milk (*kinampay* and RC 2000). The formulation with 6.25% sugar was deemed to be the best as recommended by the sensory panelists.

2.4.2.1.5 Sour starch and flour from sweet potato

Starch is one of the most important contents in the food industry. Fermentation of sweet potato flour using a consortia of microorganisms (*Lb. cellobiosus, Streptococcus lactis* and *Corynebacterium* sp.) enhanced the pasting temperature, solubility as well as reduced the swelling (Panda and Ray 2016).

2.4.2.2 Acetic acid fermentation

Acetic acid fermentation is a process in which sugar or starch rich materials are fermented to alcohol and subsequent fermentation of alcohol to acetic acid via microbial oxidation. When sweet potato is taken as a substrate, breakdown of starch to sugar is important. After this step is reached, sugars are fermented to ethyl alcohol by *Saccharomyces* sp. In the final step the ethanol produced is oxidised to acetic acid by *Acetobacter* sp. The starting raw materials vary from one region to another for the generation of acetic acid. The prime objective behind the production of acetic acid is to prepare vinegar from it.

2.4.2.2.1 Sweet potato vinegar

Vinegar is produced through reducing the strength of acetic acid by diluting it with water. Finished vinegar should have at least 4.3 g acetic acid/100 ml (Woolfe 1992). In Japan, innovative vinegar was prepared by using purple sweet potato as substrate. The vinegar was red in colour, which showed higher antioxidant activities as compared to other standard white and black vinegars. The high antioxidative property of the vinegar was attributed to the presence of 6-O-(E)-caffeoyl-(2-0-β-d-glucopyranosyl)-α-d-glucopyranose or caffeoylsophorose. The presence of caffeoylsophorose was determined in the vinegar with the support of HPLC, mass spectrometry and NMR (Terahara et al. 2003).

2.4.3 *Yeast fermentation-potential to make alcoholic beverages*

Yeast is a unicellular fungus which reproduces asexually by budding or fission. Yeasts are extensively available in all natural spheres and are mainly dominant amongst the microflora of orchards, vineyards and in other sugar rich fruits. The prime role of yeast in the food industry is alcohol production, texture enhancement, acidification and antimicrobial compound production for preservation and storage. Some species of yeasts are capable of producing hydrolytic enzymes such as amylase, lipase, cellulase and xylanase for the improved digestibility of food. Yeast is known to prevent the growth of pathogenic bacteria and spoilage microbes by generating antimicrobial agents, organic acids and hydrogen peroxide. Yeast genera associated with fermented foods and beverages are *Saccharomyces, Zygosaccharomyces, Schizosaccharomyces, Debaryomyces, Rhodotorula*, etc. For the production of conventional beer, *S. cerevisiae* is the most used and studied species. However, the starters are divided into top fermenting yeasts and bottom fermenting yeasts (Capece et al. 2018). *S. cerevisiae* is known as the top fermenting yeast used in the production of ale, stout and porter. It is also observed that many of the top fermenting yeast species found in nature are hybrids. It has been reported that one fourth of the top fermenting yeast strains in Belgian Trappist Beers are mostly hybrids of *S. cerevisiae* and *S. Kudriavzevii* (Gonzalez et al. 2008). *S. uvarum*, formerly known as *S. carlsbergenesis* is mostly used as a bottom fermenting yeast. Nowadays *S. pastorianus* is employed for the production of lager beers. Lager beer has a market stake of 90%. Lager yeasts can be divided into two types. The lager yeast can be divided into two groups, 1: Saaz and 2: Frohberg. Saaz is hybrid of *S. cerevisiae* and *S. eubayanus*, and Frohberg is further hybridized with *S. cerevisiae*. The fermentation potential of Saaz strains is slower than that of Frohberg yeasts. The application of yeasts to prepare wine is an age old practice using the auto–fermentation of fruits in tribal areas, using indigenous methods. With the isolation of various strains of yeast, the efficiency among the strains was studied. As yeast contributes a major role in wine flavour (Lilly et al. 2000) various strains of yeasts are applied for wine preparation such as *S. cerevisiae* (Walker 1988), *Hanseniaspora guillermondii*, *Kloeckera apiculata* (Romano et al. 1992), *Pichia anomala* (Rojas et al. 2001), *Candida stallata* (Ciani and Maccarelli 1998), etc. Although some yeasts other than *Saccharomyces* sp. enhance the quality of wine, however they are known to quit fermenting before completion of the alcohol conversion, as they are unable to tolerate alcohol as compared to *Saccharomyces* sp. That's why *Saccharomyces* strains are widely applied for the preparation of wine and beer. Although *S. cerevisiae* is a widely accepted common yeast in fermented food and beverage industries, but *S. pombe* is preferred for beverage production in tropical zones.

2.4.4 *Sweet potato alcoholic beverages*

Being a rich source of starch and sugar, sweet potato is employed for the production of alcohol of different grades for specific usages. Sometimes sweet potato alcohol is also used in the chemical and pharmaceutical industry as well as for beverages. Sweet potato has been used for the production of alcoholic beverages in many countries through several processes and technologies with different names. For example *'masato'* is an alcoholic beverage prepared from sweet potato by the Indian ethnic groups in the Peruvian Amazon area (Austin 1985). The Chinese use sweet potato chips to produce high grade ethanol for application in beverages. Similarly, *'shochu'* is a famous traditional beverage that originated from China and is presently one of the most popular beverages in Japan. Currently, *shochu* is produced in automatic alcohol producing plants. The raw materials used for the production of *shochu* are wheat, barley, sweet potato, etc. However, 35% of the total *shochu* produced can be attributed to sweet potato as raw material. The starchy source is applied with fungal amylase for saccharification, subsequently, fermented and distilled to obtain the beverage with an alcohol content of 20–40% (v/v) (Woolfe 1992). Now days, efforts are made to blend purple sweet potato anthocyanin into the beverage to improve its functional properties.

2.4.5 *Wine*

Wine is defined as an alcoholic fermented drink prepared from grapes and other plant parts. Archaeological study reveals that the earliest production of wine took place in Georgia, dating back 8000 years and in Armenia, 6100 years ago (Keys 2003; Berkowitz 1996). Wine originated in Europe, i.e., the natural growing area of grape vines (*Vitis vinifera*). Italy, France and Spain are the largest producers of wine with production of 42.5 million hectolitre, 36.7 million hectolitre and 3645000 million hectolitre, respectively during 2017. Figure 2.4 depicts the country wise share of wine in terms of production during the year

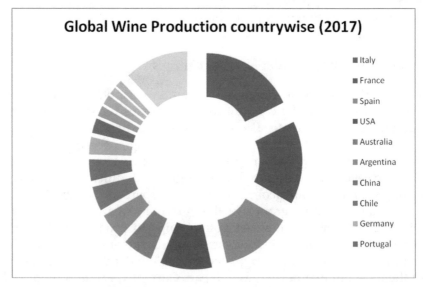

Figure 2.4. Leading wine producing countries (Source: https://www.statista.com/statistics/240638/wine-production-in-selected-countries-and-regions/).

2017 (https://www.statista.com/statistics/240638/wine-production-in-selected-countries-and-regions/). USDA (United States Department of Agriculture) nutrient database reveals the nutritional value of red table wine which has energy of 85 Kcal, carbohydrate of 2.6 g, sugar of 0.6 g, fat of 0.0 g, protein of 0.1 g and alcohol of 10.6 g per 100 g of wine. Red wine contains beneficial chemical compounds such as resveratrol, polyphenols, antioxidants, flavonoids, procyanidins, etc. these can prevent cardiac diseases, cancer, liver diseases and several other health disorders amongst its consumers. Wine is considered a prestigious drink that is served at high profile meetings and parties. Apart from Europe wine is becoming popular in other continents. In India the wine market is rising by 25–30% every year for the last decade (Gore 2008).

2.4.5.1 *Innovations in wine making*

Innovations are constantly added to the wine making process. Automation and the introduction of sensors to wineries have brought perfection in wine making. It has also made the job of wine makers simpler and brought uniformity in quality in all batches of production. Application of coculture of yeasts as a starter culture is one of the prime innovations. Coculture is applied for a reduced alcohol yield and glycerol generation in the fermentation medium; glycerol improves the smoothness in the flavour of the wine and counteracts against the burning sensation of alcohol. Contreras et al. (2014) demonstrated how the incorporation of *Metschnikowia pulcherrima* prior to inoculation of *S. cerevisiae* resulted in low alcohol production along with desired sensory property. Breeding of yeasts and application of genetic engineering has brought several desired properties in the resultant strain such as the property to produce reduced H_2S and resistance to killer factors (Jackson 2017). Application of several important technologies and equipment such as IR spectroscopy, PCR, qPCR, etc., have provided ensured quality assurance to the final product.

Innovation in wine presentation is mainly about the new packaging technology. Traditional wine bottles are made up of glass and sealed using cork, but in the present day some wines are packed in cans (single serving pack), tetra–packs, recyclable glasses, etc. Some reputed wineries are also selling their products in a box or a bag where the wine remains fresh up to 60 to 90 days after the first drink is dispensed. The wine box system prevents the oxygen from reacting with the wine.

2.4.5.2 *Tropical fruit wine*

As grape cultivation is confined to a particular climatic condition, it is not possible to meet the global wine demand only through grapes. Furthermore grapes are prone to various diseases in the tropical and subtropical regions which are a hindrance in viticulture and wineries in temperate climate. Tropical fruits are found to have a high content of fermentable sugars, which make them suitable for fermentation. For the reasons given above tropical fruit wines therefore have a lot of scope in gaining popularity (Akubor et al. 2003; Mohanty et al. 2006). A numbers of studies have been conducted on tropical fruit wine/ beverages to explain the biochemical parameters for their conversion from fruits to wine.

Banana and plantain (*Musa* spp.) were fermented to wine. The process involved the removal of peels, mixing of the banana pulp with water in a ratio of 2:1, heating it under agitation for 1 hr at 90°C and further fermentation was carried out using a starter culture of *S.cerevisiae*. The wine obtained was clear with an ethyl alcohol concentration of 8.5% (v/v). The final yield was 620–890 ml/kg (Alves 1999). Akubor et al. (2003) demonstrated the fermentation of banana juice to wine. The wine had ethanol content of 5% (v/v); protein, 0.04%, TSS of 4.8°Brix, TA, 0.85% and vitamin C of 1.4 mg/100 ml. Organoleptic analysis reported that the banana wine was accepted by the panelists and no significant difference (p > 0.05) was observed between the developed wine and a standard wine in taste, clarity and overall acceptability. Cashew apple (*Anacardium occidentale* L.) a good resource of vitamin C (3–6 times higher than that of orange juice) has been fermented to make wine. The fruit juice of cashew apple was fermented to obtain the wine which was a light yellow in colour and had the following compositions: TSS, 2°Brix; reducing sugar, 0.9 g/100 ml; titratable acidity, 1.21 g tartaric acid/100 ml; pH, 2.92 and ethanol of 5% concentration. The compositions showed a similarity with a reference wine. Cashew wine overall was acceptable in a comparative study with a standard grape wine (Mohanty et al. 2006). Mango (*Mangifera indica* L.) juice was also fermented to create wine. Reddy et al. (2005)

demonstrated the bio-conversion of six varieties of mangoes to wine. The proximate composition of the mango wines were alcohol, 6.5–8.5; titratable acidity, 0.60–0.82; pH, 3.6–4.0 and residual sugars, 2–2.5% (w/v). The biochemical property of mango wine depicts the similarity of aromatic constituents with that of grape wine.

Similarly tropical fruit wine has been successfully produced from other tropical fruits like litchi (*Litchi chinensis* Sonn.), jamun (*Syzgium cumini* L.) and soursop (*Annona muricata* L.). Wine developed from the concentrate of litchi pulp at optimum condition had ethyl alcohol concentration of 11.60% (v/v), total esters of 92 mg/l, total aldehydes of 124 mg/l and 0.78% (v/v) of titratable acidity. The litchi wine produced was clean, with light amber colour, exotic fragrance of natural litchi fruit and a pleasant wine flavour (Singh and Kaur 2009). Similarly in another study litchi fruits (*Litchi chinensis* Sonn. var. Shahi) were fermented by using *S.cerevisiae* to produce wine. The wine had the following compositions: TSS (°Brix), 2.80; reducing sugar (g/100 ml), 8.00; titratable acidity (g tartaric acid/100 ml), 0.59; pH, 3.92; phenol (g/100 ml), 0.22; tannin (mg/100 ml), 0.72; lactic acid (mg/100 ml), 0.38; ethanol (%), 6.50 (Kumar et al. 2008).

Wine was prepared by successful fermentation of jamun (*Syzgium cumini* L.) fruits through the application of *S. cerevisiae*. Jamun wine had an acidic flavour, with high tannin (1.7 mg/100 ml) content and mildly alcoholic (6%). Although the trained panelists recommended the acceptance of the jamun wine, but significant variations (P<0.05) were observed between the jamun wine and the reference wine, especially in flavour. This difference might be due to the high tannin content in the jamun wine. The authors claimed that consuming jamun wine could inhibit diabetes and bleeding piles (Chowdhury and Ray 2007).

Soursops (*Annona muricata* L.) were fermented to wine by three different types of fermentation. The authors revealed that the fermentation was at its most efficient by fermenting pasteurized soursop juice and indigenous yeast (Okigbo et al. 2009).

Tendu (*Diospyros melanoxylon* L.), an underutilized fruit in Asia was used to prepare wine (Sahu et al. 2013). *Saccharomyces cerevisiae* was employed for the fermentation process. The wine had β-carotene as a constituent and the alcohol content was 6.8% (v/v). Similarly, fruit wine was prepared by fermenting sapota (*Achras sapota* Linn.) fruit juice with wine yeast (Panda et al. 2014a). The wine had a good acceptance in terms of organoleptic attributes and the alcohol content was 8.23% (v/v). Further, titratable acidity of the wine was, 1.29 g tartaric acid/100 ml and the pH was around 3. The total phenolic content was recorded to be 0.21 g/100 ml. Infrared studies of the wine could identify alcohols, phenols, anhydrides, amides, esters and alkenes. Another study was conducted by Panda et al. (2014b) for the development of a *β*-carotene rich wine from bael (*Aegle marmelos* L.) fruits. The wine had *β*-carotene content of 33 mg/100 ml, ascorbic acid content of 80 mg/100 ml, lactic acid content of 0.64 mg/100 ml and ethyl alcohol concentration of 7.87% (v/v). Panda et al. (2016a) have demonstrated the manufacturing of jackfruit (*Artocarpus heterophyllus* L.) wine [ethanol content, 8.23% (v/v)] by fermenting the extracted juice from the pulps with *S. cerevisiae*. The wine had an exotic and unique flavour, accepted by the sensory panelists.

2.4.6 *Sweet potato as a source for fermentation into wine*

Sweet potato is rich in *β*-carotene (0–20000 μg/fresh roots), which is much higher when compared to other fruits such as banana (200 μg/ fresh fruit), guava (170–330 μg/fresh fruit), mango (335–13000 μg/ fresh ripened fruit) and papaya (205–1500 μg/fresh fruit) (Woolfe 1992). *β*-carotene, precursor of vitamin A is an essential content of fruit which improves the food value and importance of the fruit. Some varieties of sweet potato are rich in anthocyanin pigments similar to that of fruits such as grapes. Anthocyanin concentration varies from 20–6000 mg/kg fw in different coloured fruits and vegetables. (Wrolstad 2000). Anthocyanin derived from purple sweet potato shows more stability as compared to those obtained from other fruits and vegetables such as strawberry, grapes, etc. (Lu et al. 2010). In another study, conducted by Teow et al. (2007) it was observed that there is a significant variation (17–531 mg/kg roots) in the anthocyanin content of 19 selected cultivars of purple sweet potato. One genotype of purple sweet potato was found to have an anthocyanin content as high as 1820 mg/kg fw (Cevallos–Casals and Cisneros–Zevallos 2003). Purple sweet potato anthocyanin pigments show more stability to light exposure and temperature as the pigments are found in acylated forms (Hayashi et al. 1996). Sweet potato contains thiamine (0.09 mg/fresh roots) which is higher as compared to peaches (0.02 mg/fresh fruits), mango (0.03 mg/ fresh ripened fruits) and coconut (0.03 mg/fresh fruits). Similarly, sweet potato contains riboflavin (0.03 mg/fresh roots), niacin (0.6 mg/fresh roots), pantothenic acid (0.59 mg/fresh roots), pyridoxine (0.26 mg/fresh roots), folic acid (14 µg/fresh roots) and ascorbic acid (24 mg/fresh roots) in comparable quantities as that of fruits like apricots, banana, coconut, guava, mango, papaya, peaches, raisins, etc. (Woolfe 1992). The mineral content of sweet potato namely calcium (34 mg/fresh roots), potassium (46 mg/fresh roots) and iron (0.7 mg/fresh roots) are comparable to that of other fruits (Woolfe 1992). Apart from all of the above fruit like contents, sweet potato is rich in starch (60–70% of dry matter). Sugars like glucose, fructose, sucrose and maltose are important constituents of sweet potato. The total sugar of sweet potato varies from 0.38% to 5.64% depending upon the cultivar.

Starch in the sweet potato roots can be saccharified into fermentable sugars. The reducing sugars present in the sweet potato roots can directly be utilised by the yeast strains. It can be concluded that due to the fruit like quality and the presence of starch and sugar sweet potato possesses potential as a worthy raw material for the preparation of wine.

2.4.6.1 *Sweet potato wine*

A number of alcoholic beverages have been prepared by the fermentation of sweet potato mash. Sweet potato wine is prepared indigenously by various tribes of Asia, Africa and South America and households using indigenous methods. The process of wine making involves the peeling and boiling of sweet potato, addition of minced raisins and sugar and pectic enzymes, an addition of yeast for fermentation, after the completion of 4–7 days, fermentation is stopped and the wine is allowed for racking and bottling (Nelson and Elevitch 2011). Another wine like ethanolic drink called *masato* is traditionally prepared from sweet potato by certain ethnic groups of the Amazon area. Occasionally *masato* is formulated from purple fleshed or orange fleshed sweet potato in order to obtain the drink with a special colour.

2.4.7 Herbal/medicinal wine

Herbs and spices are traditionally used as additives in food and beverages for flavour, aroma, medicinal and preservative properties (Beuchat 1994; Cutler 1995). These herbs are applied for several purposes, some herbs are reported to inhibit the growth of microorganisms through their bactericidal property (Beuchat and Golden 1989) others impart a functional property to the food by incorporating essential phytochemicals, vitamins and other bioactive compounds. Many researchers have reported that the preparations of wine or wine like alcoholic beverages are using herbs as substrates or as adjuncts. Soni et al. (2009) illustrated the preparation of wine from the Indian gooseberry (*Emblica officinalis*) fruits that possess pharmaceutical properties such as carminative, anti–diarrhoeal, anti–haemorrhagic and anti–anaemic. *Aloe vera* is commonly applied in several health therapies and has been used as a substrate in the preparation of wine (Pongparnchedtha and Suwanvisolkij 2011).

Leaves of tea (*Cammellia sinensis*) have been used for alcoholic fermentation. Infused tea leaves were fermented with wine yeast and the resultant drink was accepted with both organoleptic and functional attributes (Aroyeun et al. 2005). Bhat and Moskovitz (2009) have demonstrated the preparation of a herbal tea by conglomerating an appropriate combination of African herbs. The authors claim the property of the tea includes the ability to inhibit arthritis, constipation, cardiac disease, etc. (Bhat and Moskovitz 2009). Similarly wine has been prepared from jamun (*Syzgium cumini* L.), another rich source of antioxidants. The authors claim jamun has similar anti–diabetic and anti–piles properties as wine (Chowdhury and Ray 2007).

2.4.8 Beer

Beer is the most commonly drunk beverage (European Beer Guide 2006) and presumably the most ancient (Rudgley 1993) alcoholic drink. In popularity it stands in third position amongst drinks after water and tea. Cereal grains, generally barley are often used for the preparation of beer; the prime processes involved in the preparation of beer are the malting of grains, and fermentation followed by carbonation. Keeping in view of the cheaper cost of maize and rice, the sugars derived from the sources are used as adjuncts. Hops are used in the preparation of beer for flavour, aroma and its natural bactericidal properties. Beer prepared by the fermentation of *S. cerevisiae* is known as top fermented beer while beer prepared by the fermentation of *S. uvarum* is known as bottom fermented beer. Bottom fermented beers are further classified into Pilsener beer, Dortmund beer, Munich beer and Weiss beer whereas top fermented beers are further classified into ale, porter and stout. Global beer production during 2017 was 1909 million hectoliters. China ranks top in beer production (397.8 million hectolitres) followed by the USA (217.7 million hectolitres) and Brazil (140 million hectolitres). Beer production of leading countries is presented in Fig. 2.5.

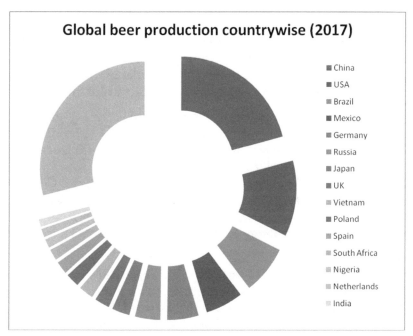

Figure 2.5. Leading beer producing countries (Source: https://www.kirinholdings.co.jp/english/news/2018/0809_01.html).

2.4.9 Innovative beer

Craft breweries are becoming more accepted as they produce small batches of beer with an emphasis on distinguished quality, flavour and brewing technique. In conjugation with craft brewery, the distillation of spirits is also promoted on the same site. This concept is known as brewstillery. A brewstillery accommodates both a brewery and distillery on one site as same units or partnering units. Sometimes two separate units of brewery and distillery are established as both can't be installed on one site due to local regulations. In the USA around 1300 brewstilleries are being run (Russell and Kellershohn 2018).

Unlike traditional beers, several alterations either in ingredients or in production technology are carried out to develop the unique products preferred by the consumers. Sale of low-alcohol or alcohol-free beer is on the rise; more than one fourth of beers produced in China are low alcoholic (alcohol content: less than 3.5%). Similarly, nonalcoholic beers are produced in two different processes. In the first type, the fermentation is brought to an end prior to the production of ethanol but after the generation of flavouring compounds and in the second process a high temperature is applied for the removal of alcohol, however, the drawback lies in degradation of the flavouring compounds resulting in a poorer flavour and aroma of the beer.

Fruit beers are prepared by taking different fruits either as a substrate or as a adjunct. Hundreds of years ago fruits were used to produce beers of the Belgian lambic styles. Some beer producers blend fruit extracts in the finished beer instead of fermenting the fruits along with the beer. A beer was formulated in 2009 by RJ Rockers Brewing Company, USA which optimized a beer in which peaches are fermented with unfiltered wheat ale. Similarly, *fruli* is a product of Belgium which is formulated by the blending of wheat beer (70%) and fruit juice (30%).

Beer making is practised traditionally from different substrates in various countries. *'Talla'* is an Ethiopian beer which acquires sonic smoke flavour from the toasted, milled and boiled cereal grains. *'Bussa'* is an acidic beer and is consumed by several ethnic communites of Kenya. The beer is prepared from millet malt powder and roasted maize flour dough. *'Merissa'* is a sour Sudanese beer with alcohol content of 6% (v/v). The beer is prepared by the fermentation of both malted and unmalted sorghum beer. The traditional beer of Western Africa and Nigeria, i.e., *'Buru-kutu'* is prepared by the addition of cassava flour (*gari*).

2.4.10 Hybrid alcoholic drinks

Russell and Kellershohn (2018) have described in details the various innovations in wine, beer and spirits. Various non-traditional combinations are currently getting more attention among their consumers. For example Speers are spirit beers and Spiders are spirit ciders. Similarly, preformulated cocktails in a ready to drink form are the new drinks preferred by the young generations rather than the traditional wine, beer and spirits.

Wine–beer hybrids are also a new concept that some breweries have started following (Peters 2018). In this hybrid, a particular quantity of the grape juice/must is blended with beer and further fermented using wine yeast. Bruery Terreux, USA produces a beverage 'confession' which is neither a wine nor a beer. Sour Ale is blended with grapes and further fermented to produce a unique beverage with 9.4% (v/v) alcohol content. Likewise, another company Allagash Brewing of Portland developed a product 'Victoria Ale' in which 500 pounds of grapes in crushed form is blended with 2000 pounds of Pilsner malt. The produce has a golden colour with a fruity aroma and an alcohol concentration of 7.3% (v/v).

2.4.11 Sweet potato beer

Similarly a beer like product preparation has been attempted by taking sweet potato flour as substrate. The Koedo brewery of Kawagoe in Japan is known to produce beer by taking the roasted local sweet potatoes since 1996. The alcohol content of the beer remains around 7% (Odebode et al. 2008). Sweet potato has been incorporated along with barley sprouts to enhance the hydrolysis pattern of starch in making of a beer like rice beverage. The incorporation of sweet potato enhanced the sweetness, flavour and improved preference in rice beverage as the enzymes of sweet potato are competent and thermostable rather than that of barley. An indigenous beer like rice beverage named *miki* was reformulated in Japan by using the uncooked sweet potato root pulp as a saccharifying agent in different combinations (Teramoto et al. 1998).

In another study conducted by Etim and Ektokakpan (1992), the replacement of 20% sorghum malt by sweet potato flour was proved advantageous in the development of sorghum beer. The resultant beer prepared with extra sweet potato flour showed low levels of free alpha amino nitrogen for which the beer had more clarity as compared to the whole sorghum beer. Further, wort prepared with a combination of sweet potato flour showed the presence of (1→3, 1→4) β–glucanase which was not observed in whole sorghum wort (Etim and Ektokakpan 1992). Food products obtained through the fermentation of sweet potato are presented in Fig. 2.6.

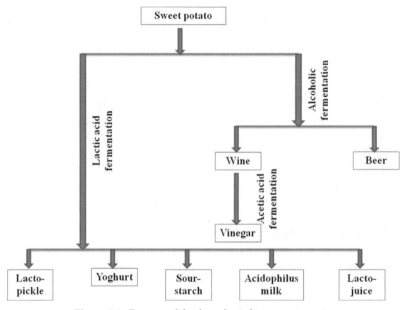

Figure 2.6. Fermented food products from sweet potato

2.4.12 *Fermentation of sweet potato for bio–commodities*

Sweet potato is also used as a substrate for the production of economically important biotechnological products. Although there are many products to be covered but in this section only on the important bio-commodities currently in use have been focused on. We have described some of the important enzymes, organic acids and biofuels of commercial significance.

2.4.12.1 *Enzymes production from sweet potato*

Microorganisms are used for the production of enzymes from cheaper substrates. Fruit and vegetable wastes have been considered as a potential substrate for the production of valuable enzymes such as amylase, pectinase, protease, cellulase, tannase, etc. (Panda et al. 2016). These enzymes have numerous applications in industries including food, textile and pharmaceutical. Both solid state fermentation (SSF) and submerged fermentation are adapted for the production of microbial enzymes from sweet potato and its derivatives as a substrate. The important microbial enzymes produced from sweet potato are amylase and cellulase. Amylases are used for the degradation of starch molecules to simpler sugars such as glucose, fructose, maltose, etc. It has applications in many industries such as food and beverage manufacturing units, textile industries and pharmaceuticals. Mostly bacteria and fungi are harnessed for the production of amylase. The production of microbial amylase is usually conducted in cheaper substrates such as cassava, sweet potato, cereal bran or their derivatives. Several findings have also suggested sweet potato starch or peel as a superior substrate for the production of amylase as compared to other substrates or their derivatives. In a study, three fungi (*Aspergillus* sp., *Mucor* sp., and *Rhizopus* sp.) were used for the optimization of the production of amylase. The optimum temperatures for production were, 45°C for α-amylase and 30°C for glucoamylase. The highest enzyme activity was observed at 120 h, pH 4 with sweet potato starch as a carbon source (Matthias 2013). Recently, Pereira et al. (2017) have demonstrated amylase production from the sweet potato peel. The optimal parameters for high production of amylase (214.28 U per ml) were 6.5 days of cultivation period, pH 6, and 2% sweet potato peel as substrate. Ubalua (2014) have used sweet potato starch for the production of amylase and it recorded that the highest glucoamylase activity was 9.40 U/mg at 40°C and pH 4.5. An interesting study was conducted on the production of α-amylase by using different agro-wastes (cocoyam waste, sweet potato waste, wheat waste and plantain peel). Sweet potato waste (2 g) with NH_4Cl as nitrogen source, 10% innoculum concentration, incubation period of 96 h, pH 5 and temperature of 50°C were the conditions for maximum amylase production (Ifeoluwa 2017). A similar study was also conducted by Yamunarani et al. (2010) to optimize amylase production from *Aspergillus oryzae* in submerged fermentation. Rice bran, wheat bran and sweet potato were used as the substrate for the production. The authors conceded that the maximum amylase yield was observed with a substrate having 2% of sweet potato; pH 6.5 and temperature of 35°C. Another study used varieties of agricultural by-products to observe the production of amylase. Solid state fermentation was conducted using *Bacillus licheniformis*. Out of the several by-products (flour of wheat, barley corn, gram, husk of moong, Arhar, oil cake of mustard and coconut oil, peel of banana, potato and sweet potato, soybean hull, bran of rice and wheat, and sugarcane baggase), used for amylase production, sweet potato peel showed enzyme activity of 149.77 U/ml whereas wheat bran had the highest titer of amylase (154.17 U/ml) (Singh et al. 2014).

Ogbonna et al. (2018) have described the production of cellulase and amylase of five fungal genera (*Aspergillus* I and II, *Penicillium* sp., *Rhizopus* sp. and *Trichoderma* sp.), isolated from rotten cassava, cocoyam, white yam, and sweet potato roots. *Aspergillus* sp. type I showed the maximum glucoamylase units, i.e., 572.16 ± 7.92 µmole/g from rice grains and 482.70 ± 2.00 µmole/g from sweet potato flour mixed with 20% rice husk.

2.4.12.2 *Organic acids production from sweet potato*

Organic acids such as citric acid, lactic acid and acetic acid possess a huge stake in the market of fermented products. Further, it beholds the third largest category among the bio-products. Organic acids are understood as weak acids and they do not dissociate completely in water. Some species of bacteria as well as fungi are commercially exploited for the production of organic acids. Amongst bacteria, *Bacillus* sp., *Lactobacillus* sp. and *Streptococcus* sp. are mostly used for the production of organic acids whereas *Aspergillus* sp. and *Penicillium* sp. dominate the production of fungal organic acids. Organic acids have applications in food and feed processing, nutrition, oil and gas stimulation units, etc. (Panda et al. 2016). Acetic acid has been described in the Section 2.4.2.2 of Chapter 2. Studies have been conducted to establish and improve the production of organic acids from sweet potato and its derivatives as substrate. Sardar et al. (2007) have depicted the production of citric acid by *Aspergillus niger*. According to the study, sweet potato starch was found to be more suitable for the production of citric acid as compared to maize starch. The kinetic parameters in fermentation, $Y_{p/x}$ (1.96 g/g cells), Q_p (0.13 g/l/h) and q_p (0.014 g/g/h) were 8–10 times more with sweet potato starch as substrate rather than the starch obtained from maize. The optimized parameters for the highest production of citric acid (28.40 g/l) were sugar, 150 g/l; incubation period, 168 h and initial pH, 6.0. A two step hydrolysis study [bacterial α-amylase (6.4 units/mL) + fungal glucoamylase (789.6 units/mL)] was conducted by using sweet potato starch for the production of sweet potato hydrolysate (Betiku and Adesina 2013). The optimal conditions for the generation of the highest sweet potato hydrolysate (241.92 g/l) was temperature, 52°C, incubation period, 44 min and pH 4.5. Further, the same sweet potato hydrolysate along with di-ammonium hydrogen phosphate, potassium dihydrogen phosphate and magnesium sulfate was used as a substrate for the production of citric acid. *A. niger* was employed for the production of citric acid and it was concluded that the optimum citric acid production (86 g/l) was obtained on the 8th day of fermentation. A bulk amount of agro-waste is generated in vegetable canning industry; similarly, the sweet potato canning industry produces huge amount of waste, i.e., around 30% of the raw material. Industrial sweet potato waste has been used for the production of lactic acid (Pagana 2012). The waste was pretreated by 80 U glucoamylase/100 g sweet potato waste at 35°C for 24 hr resulting in a yield of more than 95% of glucose conversion. *Lactobacillus rhamnosus* was further applied for the production of lactic acid. The highest lactic acid production (10 g/l) was obtained at pH 5 and 72 hours. Similarly, Adthalungrong et al. (2014) have elucidated the production of lactic acid from sweet potato. Like other studies, sweet potato was subjected to liquefaction and saccharification by using α-amylase and glucoamylase to obtain sweet potato hydrolysate. The fermentation medium was developed by using the sweet potato hydrolysate along with peptone, a meat extract, winery waste and other inorganic chemicals (Tween 80, $(NH_4)_2C_6H_6O_7$, CH_3COONa, $MgSO_4.7H2O$, 0.05 g/L $MnSO_4.H_2O$ and 60 g/L $CaCO_3$) in appropriate proportions. *Lactobacillus casei* was employed for the production of lactic acid and optimized parameters were obtained from the response surface methodology. Fermentation was carried out for 72 hr with reducing sugar, 117 g/l; $CaCO_3$, 56 g/l; winery waste, 16 g/l; $MnSO_4$. H_2O, 0.064 g/l resulting in the optimum production of lactic acid (88.99% conversion). Akoetey (2015) has produced lactic acid directly from starch instead of the dual steps, i.e., saccharification of starch to glucose and then fermentation of glucose to lactic acid. Amylolytic lactic acid bacteria, *Lactobacillus amylovorus* was applied for the direct conversion of sweet potato starch to lactic acid and adaptability of the bacteria to the rising concentration of the starch was also studied. It was observed that under controlled acidic condition of the fermentation medium a large amount of lactic acid was produced. The adapted bacteria could produce 11.80 g/l of lactic acid in the medium as compared to 6.35 g/l of lactic acid produced by the un-adapted bacteria in the medium. It was further observed that starch to lactic acid conversion of 80% was achieved during the fermentation process by the adapted *L. amylovorus*. Malic acid is another important organic acid which has applications in food, beverage and pharmaceutical industries. Sweet potato hydrolysate was used as the prime carbon source (120 g/l) along with a yeast extract (5 g/l) and citrate (7.5 g/l) for the fermentation using immobilized *Aureobasidium pullulans* cells in an aerobic fibrous bed bioreactor system. The highest poly malic acid production was 57.5 g/l which was 30.6% more than that of free-cell fermentation (Zan and Zou 2013). Adesina et al. (2014) have studied the production of oxalic acid by using sweet potato hydrolysate as a substrate and *Aspergillus niger* for fermentation. Optimized oxalic acid production (103.26 g/l) was observed at 9 days and pH 6 and sweet potato hydrolysate of 149.97 g/l in fermentation medium.

2.4.12.3 *Biofuel production from sweet potato*

In view of the fast depletion of fossil fuels, alternative energy sources are being explored in all parts of the world. Biological renewable energy sources such as bioethanol, biogas and bio-hydrogen have been proved to be capable to partially replace fossil fuels (Panda et al. 2018). Bioethanol and biogas are clean fuels and have been commercially adapted in many countries. Sweet potato and its wastes have been used for the production of such fuels through the application of fermentation technology.

2.4.12.3.1 Bioethanol production from sweet potato

Raw sweet potato was applied for the production of bioethanol. Optimization was conducted for each step towards the highest yield of bioethanol (Kumar et al. 2014). The different steps include liquefaction, saccharification and fermentation. The optimal conditions for the highest conversion of starch using α-amylase was 104–105°C, 0.15% (v/w) α-amylase (300 U/ml), 30 g dry weight sweet potato mash per 100 ml distilled water. Similarly, for saccharification, the highest glucose was generated at 0.25% v/w amyloglucosidase (300 U/ml), at pH 5 and 60°C. Subsequently, the utmost bioethanol yield, i.e., 7.95% (v/v) was obtained after 48 h of fermentation, pH 6 and inoculum size of 10%. Bioethanol was produced by simultaneous saccharification and fermentation of sweet potato flour and roots (Ray and Naskar 2008). Prior to saccharification, the substrates (flour and root) were treated with thermostable α-amylase, termamyl (Novozyme, Denmark) at 90°C for one hour. Further, fermentation was conducted by thermo-tolerant yeast, *Saccharomyces cerevisiae* and during the same period simultaneously glucoamylase was incorporated to the fermentation medium. It was observed that the ethanol yield of 258 g/kg and 96 g/kg was achieved by the simultaneous saccharification process of sweet potato flour and roots, respectively. Bioethanol is produced from sweet potato by a co-culture method, in which an amylolytic fungi, *Trichoderma* sp. is used along with *Saccharomyces cerevisiae* in a ratio of 1:4 (Swain et al. 2013). Solid state fermentation was done in different conditions of moisture, pH, temperature and nitrogen source. Highest ethanol production (172 g/kg) was observed in a medium with 80% moisture, 0.2% $(NH_4)_2SO_4$, pH 5, 10% inoculum size and temperature of 30°C at 72 hr. In another study, saccharified sweet potato flour was fermented by a co-culture of *S. cerevisiae* and *Pichia* sp. Physicochemical parameters were optimized for the fermentation process using one variable at a time and response surface methodology. As observed in the one variable at a time method, the highest ethanol yield was 127.2 g/kg of sweet potato flour at pH 5 and incubation time of 72 hr at 30°C; whereas response surface methodology could enhance the yield to 138.6 g/kg, which is around 8% higher than that of the one variable at a time methodology (Dash et al. 2017). Wang et al. (2016)

have applied cellulase and pectinase to the sweet potato residues for the liberation of glucose and subsequent fermentation to bioethanol by yeast. It was observed that 153.46 g/l glucose was produced from high gravity sweet potato residues by using cellulase and 168.13 g/l glucose was obtained from a mixture of cellulase and pectinase. It was interpreted that 209.62 g and 225.71 g of ethanol was obtained by the two different enzymatic pretreatments, i.e., application of cellulase individually and a mixture of cellulase and pectinase, respectively. Lareo et al. (2013) have established that the drying of sweet potato had no effect in the saccharification for the release of glucose as well as on the yield of bioethanol through alcoholic fermentation. It can also be recorded that use of sweet potato as a raw material for ethanol production is feasible; 4800 l of ethanol can be produced from sweet potato produced per hectare.

2.4.12.3.2 Biogas and biohydrogen production from sweet potato

Sweet potato has also been harnessed as a substrate for the production of biogas and biohydrogen. Kobayashi et al. (2014) have presented the digestion performance and microbial community during the complete scale methane fermentation of stillage of the Shochu production unit. The annual stillage production in Kyushu province of Japan is as high as 8×10^5 tons. Fixed bed reactors of capacity 283 m^3 (eight reactors) were supplemented with Ni^{2+} and Co^{2+} and were operated at a high COD (Chemical Oxygen Demand) of 14 kg/m^3.day. The dominant microorganisms in the thermophilic fermentation for the generation of methane were from the phyla Firmicutes along with Bacteroidetes. *Methanosarcina thermophila* as well as *Methanothermobacter crinale* were the dominant microbes among the methanogens. Biogas production was estimated by *in vitro* fermentation (buffered and sieved goat's rumen liquor) of sweet potato and wild coco yam peels. In the 24th hour of fermentation, 42.5 ml of methane and 35 ml of carbon dioxide was produced from fermented sweet potato peel whereas in the case of coco yam peels the yield was lower, i.e., 39.5 ml methane and 32.5 ml carbon dioxide (Adeyosoye et al. 2010). Sweet potato starch was employed for the production of biohydrogen and the highest yield of 3501 ml/l in the optimum conditions (initial pH, 6.05, sweet potato starch concentration of 27.63 g/l and $FeSO_4$, 63.17 mg/ml). The seed sludge, collected from an anaerobic sludge blanket waste water treatment plant was thermally treated in a hot air oven at 105°C for inhibition of probable bacteria that could consume the hydrogen generated in the fermenter (Vi et al. 2017).

2.4.12.4 *Other important products produced by bioprocessing of sweet potato*

Other important products that have been developed using sweet potato as a substrate are single cell proteins, amino acids, antibiotics, chitosan, etc. Several studies have demonstrated the protein enrichment of sweet potato and its residues by using different microorganisms such as *Aspergillus niger, A. oryzae, Pleurotus ostreatus, Endomycopsis fibuligera, Candida utilis,* and *Trichoderma koningii* (El Sheikha and Ray 2017). Similarly, oxytetracycline has been produced by solid state fermentation of sweet potato (Yang and Yuan 1990) and its pre–treatment along with submerged fermentation (*Brevibacterium glutamicum*) of sweet potato starch produces glutamic acid and it is further converted to monosodium glutamate. In another study conducted by Nitar and Stevens (2002), improved chitosan production was observed from the mycelia of *Gongronella butleri* USDB 0201 by the fermentation of sweet potato.

Unfermented Food Products from Sweet Potato

Several unfermented products are available from sweet potatoes that are processed for human consumption. The processing includes drying, grating, mashing, slicing, blending them with other substrates, etc. Such products are produced in a specific region of the world by a particular community and these are associated with the culture and traditions of the communities. Further, some of these sweet potato unfermented products have gained popularity throughout the world and have also been commercialized. This chapter covers the preparation technology of important unfermented food products made from sweet potato.

3.1 Sweet Potato Flour and Bread

A study was conducted by Owori and Hagenimana (1998) in Lira, Uganda. The study was done for the partial replacement of conventional raw materials such as wheat flour by sweet potato in the preparation of snacks. The findings of the study claimed that it is feasible to replace 40–60% of the conventional raw materials by using either grated or boiled sweet potato. Sweet potato flour can also be used to replace 30% of wheat flour in the preparation of snacks. The incorporation of sweet potato did not significantly impact the organoleptic properties of these snacks, and were hence acceptable to the consumers; it is also important to record that the production cost was also reduced by a range of 20–64%. The reductions of the production costs are dependent on the type of the product and the sweet potato variety.

An interesting study was conducted by Nogueira et al. (2018) which demonstrated the use of β-carotene rich sweet potato in the manufacturing of bread. The experimental design was carried out by replacing 3%, 6% and 9% of refined wheat flour by β-carotene rich sweet potato flour in three different formulations along with a control sample (0% sweet potato flour). All of the other ingredients such as palm fat, salt, instant dry yeast, calcium propionate, diacetyl tartaric acid esters of mono and disaccharides, α-amylase except sucrose were the same for all the formulations. Sucrose was used differently keeping in mind reducing the sugar content of the blended flour (sweet potato flour + wheat flour). The physicochemical analysis of the different formulations at day one revealed that the specific volume (ml/g) decreased with a higher replacement (control, 3.94; 3%, 3.74; 6%, 3.69; 9%, 3.47) of wheat flour by sweet potato flour and the firmness (N), was measured by using a texture analyzer with 50 kg load cell and P/36 cylindrical aluminium probe (control, 9.61; 3%, 9.38; 6%, 7.74; 9%, 8.85). However, water activity (control, 0.939; 3%, 0.944; 6%, 0.947; 9%, 0.951) and carotenoid content (0.1656–0.4715 µg/g) showed an increasing trend. It can be conceded that the application of sweet potato flour as an adjunct is an important step towards the making of a vit-A rich functional bread. Peng et al. (2003) have provided a sequence of operating procedures for the manufacturing of sweet potato bread. The processes are as follows: (a) activation of dried yeast, (b) preparation of the primary dough, (c) fermentation of the primary dough, (d) preparation of the secondary dough, (e) fermentation of the secondary dough, (f) shaping and fermentation for the rising of the dough, and (g) baking. It is suggested that, for the manufacture of a good quality of bread, sweet potato flour should be restricted to within 30% of its content.

Similarly, Perez et al. (2017) demonstrated the application of heat treatment to flour for the manufacture of sweet potato–wheat bread. The study revealed that the heat treatment of sweet potato flour at 90°C for 20 min could improve the specific loaf volume and further gas retention of the dough increased from 1199 ml (no heat treatment, control) to 1214 ml (treatment at 90°C). Therefore the findings suggest for the heat treatment of sweet potato flour to improve the overall quality of sweet potato–wheat bread. In another study, blends of flour (sweet potato flour + wheat flour) were used to prepare bread. Five different combinations of flours were taken for the preparation of bread and further analyzed. The combinations were 100% sweet potato flour, 100% wheat flour, 5% sweet potato flour + 95% wheat flour, 10% sweet potato flour + 90% wheat flour and 15% sweet potato flour + 85% wheat flour. Physicochemical analysis revealed that no significant difference was observed in terms of crust colour, crumb holes, stability and elasticity when the proportion of the composite flour is changed up to a maximum sweet potato flour content of 15% (Oluwalana et al. 2012). Another research team attempted to replace 10%, 20% and 30% of the wheat flour that impacted on the final composition of the pan bread. The control sample (100% wheat flour) had a moisture and protein content of 22.74% and 11.6%, respectively which was found to lead to decline in the bread prepared from flour with 30% sweet potato flour + 70% wheat flour, i.e., 22.01% moisture and 9.60% protein. Contrarily, crude fibre, ash and fat were found to increase from the bread prepared from 100% wheat flour (crude fibre, 2.07%; ash, 2.40%; fat, 3.58%) to the bread prepared with 30% sweet potato flour as adjunct (crude fibre, 2.77%; ash, 3.36%; fat, 5.20%). Recently, Julianti et al. (2017) recommended an ideal composite flour consisting of 40% sweet potato flour, 40% maize starch, 19.5% soybean flour and 0.5% xanthan gum for the preparation of bread with an acceptable physical and sensory properties.

3.2 Sweet Potato Bakery Products

Preparations of sweet potato cakes were done using the flour of three different cultivars of sweet potatoes obtained locally at Bangkok, Thailand. The three varieties used for the manufacture of cakes were Man Kratai, Man Khai, and Man To Phuak. The physicochemical parameters were stable when the wheat flour was substituted with 25% of sweet potato flour. The cake prepared with 25% of Man khai variety had the highest acceptability in terms of organoleptic scores (Sripanomtanakorn and Siriboon 1999). Mu et al. (2017) have described the health potentials of purple sweet potato cakes and the cakes have been proved to be beneficial for dieto-therapy. The ingredients suggested for the formulation of purple sweet potato cake is purple sweet potato flour (90 g), mixed flour (100 g), egg (200 g), sugar (80 g) and water (15 g). The process adapted for the preparation of such cake is: (a) flour preparation (mixing of purple sweet potato flour and wheat flour), (b) whipping of eggs, (c) batter preparation, (d) filling of the molds, and (e) baking at 175–180°C for 15 min. However, the surface of the cake was observed to be dark brown in colour, perhaps due to the non-enzymatic browning but the colour of the cross section was in turn pleasing to the eye. To overcome the surface browning of the purple sweet potato cakes, the bread was baked using a microwave instead and a desirable colour both in the surface and the cross section was observed. Sweet potato cookies have been prepared by blending appropriate quantities of sweet potato to wheat flour. Sweet potato is first steamed, peeled and then mashed. Dough is prepared by mixing sweet potato mash along with milk solids, wheat flour, baking soda, flavours etc. Cookie extruding and shaping is carried out followed by baking at 230–250°C for 5–6 min (Mu et al. 2017). Similarly, sweet potato flour was blended with wheat flour in different proportions for the manufacture of cookies. Rheological studies shows that the dough prepared with 20% sweet potato flour had improved cohesiveness (0.480 g/s) as compared to 0.295 g/s in the control sample (cookie from 100% wheat flour). It was also observed that the use of an excess proportion of sweet potato flour (60% or more) during the preparation of cookies leads to a poorer organoleptic attributes of the food product. In another study conducted by Jemziya and Mahendran (2017), sweet potato (cv. Waripola Red) flour was used for the preparation of functional cookies with enhanced physical and nutritional qualities of the product. The authors claimed that replacement of 40% of the wheat flour by sweet potato flour of cultivar Waripola Red was the most acceptable formulation as per the sensory characteristics. An innovative study was conducted by using composite flour with 30% sweet potato flour and 70% wheat flour for cookie preparation. Furthermore, brewers spent grain was supplemented in different quantities (1–9% of the weight of the flour) to the blend. Biochemical analyses indicated that cookies prepared with 9% brewers spent grain had the highest content of fibre whereas the cookies prepared by fortification with 3–6% brewers spent grain was preferable amongst the sensory panelists as compared to cookies prepared from other fortified flours.

3.3 Sweet Potato Biscuits

Sweet potatoes have been used for the manufacturing of biscuits. Sweet potato granules are used along with conventional flours for the production of biscuits (Mu et al. 2017). Other supplemental ingredients are used in the manufacturing process such as vegetable oils, sugar, salt, eggs, etc. The appropriate proportion of ingredients recommended for the production of such biscuits are flour with high gluten (100 g), sweet potato granules (30 g), grease (15 g), sugar (15 g), sodium bicarbonate (1 g) and ammonium bicarbonate (0.8 g) along with other additives such as salt, flavouring agents, colourants, etc. The manufacturing process involves mixing (sweet potato granules + conventional flour + supplemental additives), dough concoction (28°C), rolling, baking (surface temperature, 220°C, bottom temperature 190°C for 9 min), cooling (room temperature) and packaging. An innovative biscuit was prepared by using unconventional composite flour prepared by blending unripe banana (cooking variety), pigeon pea and sweet potato (Adeola and Ohizua 2018). Blended flour of recommended proportion (21.67% unripe banana + 21.67 pigeon pea + 57.67% sweet potato flour) resulted in biscuits of higher preference by consumers. Furthermore, the use of this unconventional raw material enhanced the nutritional property of the biscuit (improved dietary and protein). Jumirah and Lubis (2018) have developed a technology for the development of a nutritious biscuits by using sweet potato flour and tempe flour, enriched with vitamin A from red palm oil. The biscuits of different proportions of sweet potato flour and tempe flour were found to have high a content of protein (8.84 to 14.48%), iron (4.18 to 5.59 mg), zinc (4.29 to 7.60 mg) and β-carotene (14.93 to 17.28 mg) as compared to conventional biscuits. Onabanjo and Dickson (2014) also used sweet potato and wheat flour composites in different ratios (100:0, 90:10, 70:30, 60:40 and 50:50) to prepare biscuits. The study revealed that the protein, crude fibre and fat content in the biscuits ranged from 4.50–8.92 g/100 g, 3.16–5.10 g/100 g and 10.97 g–18.93 g/100 g, respectively. The highest crude fibre was observed in the biscuit prepared with 50:50 sweet potato flour and sweet potato flour combination. During the same year another piece of research was conducted in Nigeria; sweet potato, plantain and malted sorghum were used as substrates for the preparation of biscuits (Ishiwu et al. 2014). Mixture response surface methodology was applied to model the taste and texture of the biscuits with different combinations of the flours obtained from the raw materials and it was conceded that sweet potato flour had the highest positive influence on the taste while on the other hand malted sorghum affirmatively influenced the texture. Mais (2008) has depicted the incorporation of 5–10% of sweet potato during the preparation of biscuits with an acceptable sensory property.

3.4 Sweet Potato Starch

Starch is an important polymeric carbohydrate molecule used in the food industry, paper making, pharmaceutical industry and the production of resistant starch, etc. Starch is synthesized in the plants leaves during the daytime and stored in granular forms; furthermore, it serves as a source of energy during night. Starch is obtained from various food sources such as cereals, fruits such as banana and unripe apples and tuber crops. Among cereals, maize, sorghum, wheat and rice are exploited for the extraction of starch whereas among tuber crops sweet potato and cassava are used to obtain starch. Native starch is a white coloured powder with a bland flavour and it is extracted by a wet separation technique of the ground crop. Starches of tuber crop origin are known to possess lesser amounts of lipids and proteins as compared to cereal starches. The range of lipids in cereal is 0.2–0.8% whereas in the case of root crops it ranges from 0.1–0.2%. Similarly, the protein content in cereals ranges from 0.2–0.5% and in tuber crops the variation of the same is 0.1–0.2%. Because of the low levels of protein and lipid contents in the tuber starch, the transparency of the starch paste is higher compared to cereal starch. Furthermore the higher protein and lipid interference in the cereal imparts a 'typical raw cereal flavour' in the starch of the cereal's origin, which is one of the prime drawbacks of cereal starch.

Starch granule size vary from very small (1.5–9 μm in rice and oat starch) to larger granules (around 100 μm in potato). The shape of sweet potato starch granules are round or oval and their size varies from 3–28 μm (Moorthy and Shanavas 2010). Solubilization of starch granules in water is reversible below the gelatinization temperature as the starch is stable and semi crystalline in its structure. In temperature below the gelatinization temperature the water absorption is less than 40%. Gelatinization of the starch granule is understood as the process in which the granule swells and disrupts itself resulting in the loss of the crystallinity and birefringence. The gelatinization temperature of sweet potato depends on the variety in question and the range is largely between 58–84°C. Furthermore, sweet potato starch contains one third of amylase and two third of amylopectin (Adams 2004). Amylopectin is responsible for the crystallinity of the sweet potato starch; higher the proportion of amylopectin in sweet potato, higher the gelatinization enthalpy. The gelatinization enthalpy of sweet potato starch ranges from 7.8–15.5 J/g (Mu et al. 2017). Swelling of starch granules is another important characteristic of starch. The swelling characteristic of a starch granule is presented by swelling volume. Swelling volume is estimated as the ratio of swollen granules in the sediment out of the total dry starches at a particular temperature. Several researchers have worked on the different interfaces of the swelling of sweet potato starch granules. Ahmed et al. (2010) and Moorthya (2010) revealed the swelling power of sweet potato starch granules to be 32.3–50 mg/ml at 85°C. Starch solubility is impacted by several factors such as intermolecular forces, the swelling volume as well as the interference from lipids, proteins, surfactants, salts and sugar. Researchers have conceded that the swelling and solubility of sweet potato starch is higher than that of maize starch but lower than that of the starch obtained from potato and cassava. During the storage period, the gelatinized paste becomes hazy and an insoluble white powder deposits at the bottom known as recrystallization or retrogradation. Various factors are known to influence the retrogradation of starch. For example a lower temperature and pH value of 5 to 7 triggers the recrystallization process; on the other hand anions and cations retard the retrogradation of starch. Sweet potato starch shows a low to moderate rate of recrystallization. The slower rate of recrystallization is attributed to the higher level of amylopectin as compared to amylose. For this reason, corn starch recrystallization is the fastest due to the presence of 28% amylose. Sweet potato amylose and tapioca amylose show a similar rate of recrystallization which is slower than the amylose of potato.

Starch from sweet potato is further modified to several other forms such as pre-gelatinized starch, acid modified starch, enzymatically modified starch and stabilized starch for use in food and beverage industries. An innovative approach was attempted to improve the starch yield from sweet potato by treatment of starch with citric acid (Babu et al. 2015). Along with the starch yield the water holding capacities as well as the emulsion properties were observed to improve at a higher acid concentration (range of 1–5%).

3.5 Sweet Potato Slices, Chips and French Fries

Traditionally, sweet potato slices are prepared in Africa which is meant for storage of the roots for consumption during a period of scarcity of food. In certain regions of Tanzania, sun drying of sweet potato roots is carried out for the production of *michembe* and *matobolwa*. In case of *michembe,* the roots are subjected to withering, further sliced and dried whereas *matobolwa* is prepared by boiling the roots followed by slicing and drying (Nicanuru et al. 2015). These products can be stored for 5–10 months. The traditional drying of the slices was conducted on the roof of thatched houses. But currently, with the advent of new technologies and different materials for the building of houses, alternative drying processes are adapted. Bechoff et al. (2009) demonstrated the application of different types of drying technologies to obtain dried sweet potato from a fresh orange fleshed sweet potato. The different drying processes were cross flow, greenhouse solar, and open air–sun. It was observed that hot air cross flow application to the fresh root preserved a significant amount of provitamin-A as compared to sun drying. Sometimes, the surface of the slice is coated with sugar powder for improving its taste before packaging (Mu et al. 2017).

Chips from different origins have a good market scope and are currently accepted throughout all countries and communities throughout the world. Several researchers have reported the preparation, physicochemical, and sensory property of sweet potato chips. A study was recently conducted by Caetano et al. (2018) regarding the physicochemical and sensory characteristics of sweet potato chips prepared in different cooking conditions (deep frying using canola oil, baking using oven and air frying). Three different varieties of sweet potato, i.e., cream peel and yellow fleshed, pink peel with yellow flesh and white peel and fleshed were used in the chips preparation. Chips prepared from deep fried cooking of pink peeled yellow fleshed sweet potato showed the highest acceptance among the tasting panelists. However, air fried and oven fried chips showed lower level of fat and moisture content. Mu et al. (2017) have illustrated the instant technology for the preparation of sweet potato chips. The preparation technology includes thorough cleaning and washing of sweet potato roots, peeling and slicing upto 1 cm in thickness followed by slurring, homogenization and pregelatinization. Finally, drying is carried out by a roller drier. Size of the chips can be changed by adjusting the granular mesh density. Another interesting study was conducted by Xu et al. (2012) for comparison of sweet potato chips obtained by vacuum belt drying and deep fat frying (high oleic sunflower + cotton seed oils). Four different temperatures were used for vacuum belt drying, i.e., 100°C, 120°C, 140°C and a combined heat treatment (100°C, 120°C, and 140°C). It was observed that the chips prepared by combined heat treatment were the most preferable because it showed evidence of lowest degree of hardness and high fracturability. However, the chips obtained by vacuum belt drying in lower temperatures were found to retain higher β-carotene (55.5–65.9%), whereas in case of frying or high temperature drying, 33.1–34.2% β-carotene was retained.

Currently, with the advances in various technologies many varieties of chips type of product are produced from sweet potato. Airflow puffed sweet potato chips are a special type of product in which sliced sweet potatoes are subjected to vacuum–microwave puffing, a pre-treatment step. Furthermore, the final product is obtained by processing it in an airflow explosion puffing machine. Similarly, vacuum microwave drying is another process for the manufacture of sweet potato chips. Vacuum microwave drying equipment is required to manufacture sweet potato chips of this type. Mu et al. (2017) have illustrated the conditions for the processing of sweet potato chips by this process (vacuum, 0.085 MPa; moisture of sweet potato chips before processing, 31.23%; microwave strength, 771.38 W and time of heat treatment, 39 s).

French fries are also produced from sweet potato. The variety of sweet potato impacts on the physical and sensory property of the food product. The preparation technology involves peeling, blanching, drying and frying. Frying can either be deep fat frying or vacuum frying. Deep frying is widely accepted, however, vacuum frying is advantageous in some aspects as sweet potato fries produced by vacuum frying are not easily darkened and also the absorption of oil is less. In the case of large scale production of sweet potato French frics the vacuum frying is not feasible (Button 2015). Another study was conducted by Odenigbo et al. (2012) in Canada on the preparation of French fries from five varieties of sweet potato (Ginseng Red, Beauregard, White Travis, Georgia Jet clone 2010 and Georgia Jet). It was observed that among all the varieties taken for the preparation of sweet potato French fries by deep oil frying Ginsen Red was the most suitable one as it showed the lowest oil saturation and attractive colour and texture. Truong et al. (2013) evaluated the acrylamide formation in sweet potato French fries. Acrylamide is a carcinogenic compound which is produced in thermally processed food (Maillard reaction). However, the study revealed that when the sweet potatoes were pretreated with blanching and soaking in 0.5% sodium acid pyrophosphate the acrylamide concentration was reduced by seven times in different frying conditions. At a cooking condition of 165°C for two min, the acrylamide level fell from 124 to 16.3 ng/g fw. Similarly in 3 and 5 min cooking at 165°C the fall was 255.5 to 36.9 ng/g fw and 452 to 58.3 ng/g fw, respectively.

3.6 Noodles and Pasta from Sweet Potato

Noodles are one of the most popular food products in Asia (having originated from China) and have become a cosmopolitan food over the course of time. Many studies have been conducted for the preparation of noodles from sweet potato. Ginting and Yulifianti (2015) have characterized the noodle prepared by mixture of sweet potato (orange fleshed) flour and wheat flour. Mixed flour was prepared by blending of 40% β-carotene rich sweet potato and 60% wheat flour to obtain the noodles. The noodle had 6854 µg/100 g dw of β-carotene which signifies that intake of 100 g of wet noodles can fulfill 46% of minimum RDA (recommended dietary allowance) in case of adult and 91% minimum RDA in case of children. Similarly, Adedotun et al. (2015) studied some parameters during the manufacture of noodles from sweet potato starch in a single screw cooking extruder. It was observed that the protein content of the noodle showed a linear decreasing trend with feed moisture. Further, water absorption, cooking time and solubility with water was indirectly proportional with the feed moisture which indicates that water absorption, cooking time and solubility falls with the rise in feed moisture content. The authors concluded that the noodles produced by using a barrel temperature of 110°C, screw speed of 100 rpm and feed moisture of 47.5% had the best output in terms of organoleptic property. Instant noodles which have been largely prepared by wheat flour, have been partially replaced by sweet potato flour. Taneya et al. (2014) have established that instant noodles with 70% wheat flour + 30% sweet potato flour showed an increased water absorption and enhanced volume during serving. This same formula (70% wheat flour + 30% sweet potato flour) was the most accepted amongst the sensory panelists. An interesting study has depicted the structural property of the noodles obtained from sweet potato starch and mung bean starch. The structural analysis was conducted by gel permeation chromatography, a scanning electron micrograph, differential scanning calorimetry and X-ray diffraction (Tan et al. 2006). Noodles obtained from sweet potatoes were comparatively susceptible to the attack of enzymes (amylase) and acid (HCl). Due to the lower content of amylose and higher content of amylopectin in the structure of the sweet potato noodles as compared to mung noodles, it showed a higher crystalline pattern and improved adhesiveness. Hence, more study is required to improve the structural property of the noodles obtained from sweet potato starch. Preparation of pasta from sweet potato as prime ingredient has been optimized by Singh et al. (2004) by response surface methodology. In this study the variables were sweet potato flour, soy flour, water, Arabic gum and carboxy methyl cellulose. The best quality of the pasta was determined based on the organoleptic property, solid loss and hardness. The optimized parameters for the formulation of sweet potato pasta were sweet potato flour (674 g/kg) + water (195 g/kg) + soy flour (110 g/kg) + Arabic gum (10.6 g/kg) + carboxy methyl cellulose (10.1 g/kg). Protein fortified pasta was developed by using sweet potato. Three different types of protein rich substrates, i.e., whey protein concentrate, defatted soy flour and fish powder were used to improve the protein content of the pasta. The authors claim that the pasta prepared by sweet potato and whey protein concentrate had the best quality amongst all as it showed an excellent starch–protein network. Further, it was higher in essential amino acid ratio, lowest in rapidly digested starch and had the highest resistant starch. Hence, this food, i.e., sweet potato pasta fortified with whey protein concentrate may be regarded as a low GI and high protein rich food.

3.7 Sweet Potato Purees

Prior to the processing of sweet potato for puree production, washing, peeling of roots, trimming and cutting is carried out. In modern times washing is conducted through high pressure spray washers coupled with rotating brushes. Similarly, peeling in industries is done using a thermal blast process. Furthermore, the reduction of the size of the sweet potato roots is done using a hammer mill or pulp finisher. Finally the puree is produced by steam cooking. α-amylase and β-amylase are supplemented to breakdown the starch to maltose, maltotriose, glucose and dextrins. When cooking is conducted at 70–80°C, maltose is the only sugar produced in first 10 min. A good quality of sweet and well flowing purees were obtained by applying a moderately low temperature and a longer cooking time. The puree was then packed in cans for the use by customers (Truong and Avula 2010). In another study, sweet potato puree was restructured by an incorporation of calcium alginate complex. The application of calcium/alginate improved the firmness of the puree and the gel strength of the restructured puree was the highest at the room temperature (25°C) (Fasina et al. 2003). Figueira et al. (2011) have optimized the cooking conditions for a high quality sweet potato puree (90% colour retention and 70% ascorbic acid retention) which was cooked at 90°C for seven minutes. The puree was used as an ingredient in the formulation of weaning food. Coronel et al. (2005) demonstrated an aseptic packaging of sweet potato puree, processed by a continuous flow microwave heating technology. Small-scale test (carried out in a 5 kW) reports indicated no significant change in the colour and viscosity with different thermal treatments. Further, the pilot scale experiment was conducted with a 60 kW unit and the puree was treated at 135°C, which showed no microbial counts when stored upto 90 days (room temperature). Similarly, in another study, orange fleshed sweet potato puree was used during the preparation of bread (Wanjuu et al. 2018). Bread was prepared by replacing 30–45% of wheat flour and it was concluded that the bread with sweet potato puree as ingredient showed longer shelf life period, i.e., six days as compared to four days in white bread. Puree obtained from the yellow fleshed sweet potato when blended for production of bread was found to enhance the water binding capacity and reduce the water activity of the final product.

3.8 Baby Foods from Sweet Potato

Baby food is one of the most important specialized foods. Food scientists throughout the world are focusing on the preparation of complementary baby/infant food for consumption after the baby reaches 6 months of age. Furthermore, considering the case of India, four out of ten children are not getting an appropriate nutritious diet resulting in under nutrition, stunting and wasting (http://in.one. un.org/un-priority-areas-in-india/nutrition-and-food-security/). Malnutrition is attributed to the lack of affordability of the major rural and lower income group urban Indian families to the market of available fortified infant and baby foods due to their high pricing. Hence, several research projects are being conducted to develop affordable baby foods with appropriate nutrients as per the FAO/WHO guidelines by using sweet potato like substrates. Research has been successfully translated for the societal application of the developed baby foods. For example *Amul*, a CSIR–CFTRI developed a vitamin enriched milk solid product which is a popular baby food. Similarly, Bal–Ahar, a complementary baby food has also been developed by using flour of soybean, cottonseed, peanut and milk. Sweet potato has been recommended by FAO for the preparation of complementary foods for children under two years of age (http://www.fao.org/3/am866e/am866e00.pdf). It has been instructed that the roots should be cleaned, peeled, chopped, cooked and mashed to a uniform paste for feeding the children. At six months of age a children can be introduced to sweet potato, eggs, vegetables, etc. Children of 6–8 months are recommended to be provided with sweet potato (half bowl), peanuts (one table spoon) and pumpkin leaves (one table spoon) or yellow sweet potato can also be cooked along with fish, pumpkin. For children of 9–11 months old, rice along with sweet potato is recommended. Recently, a complementary baby food was developed from β-carotene rich sweet potato in order to reduce the prevalence of vit-A deficiency in Ghana (Laryea et al. 2018). The formulation was carried out using millet and soybean flours along with sweet potato and it was found that the most acceptable (organoleptic) formula was 50% β-carotene rich sweet potato + 15% millet + 35% soybean flours. The formula was found to have a higher content of protein and β-carotene, i.e., 16.96% and 0.53 mg/100 g respectively as compared to the control sample. In Nairobi, Kenya a study was conducted by Joseah (2011) for the preparation of a baby food by using malted finger millet (40%), β-carotene rich sweet potato (10%), soybean (10%) and peanut (40%). The precooked baby food had the following composition: carbohydrate content, 53.65%; crude protein, 15.96%; crude fat content, 15.94 and total ash content, 3.45. In Uganda a study was conducted to develop a complementary baby food by using sweet potato. Two different recipes were formulated: (a) sweet potato (dried), 128 g; fish, 40 g; soy, 85 g; sunflower oil, 15 g and water 732 g and (b) sweet potato (dried), 80 g; skimmed milk, 41 g; soy, 41 g; sunflower oil, 28 g and water 810 g. In the formulation with fish vit-C and vit-E are added to restrict the oxidation of the fish oil. Further, the samples are cooked, frozen at –80°C, freeze dried and grinded for use. The total carbohydrate for formula a and b were 66 g/100 g and 55 g/100 g respectively. The protein and crude fat content of sample a and b were 20 g/100 g, 2 g/100 g and 28 g/100 g, 3.4 g/100 g respectively. Furthermore the digestibility of the formula was compared with a commercial standard baby food (Cerelac, Nestle). The digestibility of sample a, sample b and cerelac were 69.5, 67.4 and 64.9, respectively (Nandutu and Howell 2009). Amagloh and Coad (2014) have estimated the nutritional composition of infant foods, namely produced from β-carotene rich sweet potato, cream fleshed sweet potato, maize–soybean-groundnut based infant food and wheat based commercial complementary food (Cerelac, Nestle). It was recorded that all the formulas except maize–soybean–groundnut based infant foods had a quantifiable ascorbic acid content. Further, it was observed that the complementary baby food produced by using β-carotene rich sweet potato was the most accepted in terms of nutritional contents.

3.9 Sweet Potato as Source of Colourants

Consumers are becoming aware about the use of colourants, especially when food is taken into account. Currently, the consumers' preference is more inclined towards the natural colourants because of the toxicity associated with many synthetic colourants (Panda et al. 2018). Several common synthetic colourants such as allura red, brilliant blue and yellow 5, 6 are known to exhibit toxicity and some exhibit carcinogenic properties (Thomas and Adegoke 2015). For example red 40, yellow 5 and 6 contain benzidine and 4 aminobiphenyl, which have been associated with cancer. Natural colours, are also known as exempt colours are advantageous over the synthetic colours like tartrazine and indigotine in terms of health promoting potential. Many vegetables and leaves of plants contain several colourful pigments such as chlorophylls, carotenoids, and anthocyanins. Many colouring compounds are represented by bioactive pigments; β-carotene imparting yellow or orange, lycopene responsible for red colour and anthocyanin for purple or pink colour. These pigments are known for their antioxidant activity. Different fruits and vegetables contain diverse bioactive pigments of different concentrations, which can be extracted as a natural colourant for applications in the food and beverage industries. Tomato is rich in lycopene, carrot contains β-carotene and beet root, grapes and purple sweet potatoes contain anthocyanin pigments.

Several studies have been conducted for the efficient extraction of anthocyanin and β-carotene pigments from purple sweet potato and orange fleshed sweet potato, respectively. A long time ago, Kim et al. (1996) have established that the anthocyanin pigments of the purple sweet potato were properly extracted by polar solvents, whereas it was scarcely extracted by nonpolar solvents. Extraction was done by using different combinations of polar solvents, among which the solvent with 20% ethanol concentration and 0.1% citric acid was found to be the most fitting for the extraction of anthocyanin pigments. The pigment solution exhibited a change in colour during the change in pH, a bathochromic shift was observed with the rise of pH. An interesting study depicts the higher extraction of anthocyanin by using acidified electrolysed water (pH 3) in assistance with microwave baking (Lu et al. 2010). The yield was comparatively higher (250%) than the conventional ethanolic extraction. The DPPH scavenging activity of the anthocyanin extracted from microwave–acidified electrolysed water was found to be superior to the conventionally yielded anthocyanin and also had a comparatively lower cost of production. Bridgers et al. (2010) have conducted a study for efficient extraction as well as enzymatic hydrolysis of purple sweet potato. They observed that an acidified solvent of methanol could produce 16–46% higher anthocyanin as compared to non-acidified methanol. The highest anthocyanin extraction (186.1 mg cyanidin-3-glucoside/100 g fw) was observed at 80°C with 3.3% (w/v) acidified methanol. Similarly, Huang et al. (2010)

have studied the optimization parameters for extraction of anthocyanin from a variety of purple sweet potato (TNG 73). They have used three independent factors in response surface methodology for the study, percentage of HCl in ethanol, extraction time in min, and extraction temperature in °C. The optimum conditions for the anthocyanin yield were 12% HCl (1.5 N HCl in ethanol), 45 min in 50°C. Truong et al. (2012) have optimized the conditions for the highest anthocyanin yield using pressurized liquid extraction. An accelerated solvent extractor was used for the extraction of anthocyanin pigments in this process. The highest yield was observed in an acetic acid: methanol: water mixture of 7:75:18% (v/v), sample weight of < 0.5 g and 80–120°C; at a constant pressure 1500 psi. Li et al. (2013) have depicted the optimum extraction (11.6355 mg/g) of the natural anthocyanin at 60°C, 1 h of extraction time, solid: liquid :: 1:30 and acidified ethanol strength of 80%. It was further observed that Fe^{3+}, Al^{3+} and ascorbic acid enhanced the stability of the anthocyanin pigment. Anthocyanin pigment of purple sweet potato has been extracted and encapsulated using maltodextrin, further the microencapsulated anthocyanin was applied to a jelly drink. The stability of the pigment was studied in different storage conditions (light exposure at 60 watt, refrigerated and exposed to light with a lamp at 30 watt, cardboard boxes at room temperature, refrigerated at temperature of 5°C in dark). It was inferred that storage of the juice with microencapsulated anthocyanin in refrigerator and in dark was the best in terms of anthocyanin and colour retention (fall of anthocyanin, 46.03% and reduction of red colour intensity, 3.19% in 30 days). Likewise β-carotene pigments are abundant in orange fleshed sweet potato. A study was conducted to statistically optimize the extraction of β-carotene pigments from orange fleshed sweet potato, (cultivar: Beta 1) (Ginting 2013). Two factors were considered for the experiment, i.e., moisture content of the grated roots and the ratio of ethanol to acetone in the extraction solvent. Highest content of β-carotene was extracted (235.94 µg/ml) from roots with moisture content of 60.63%, and extraction solvent with ethanol to acetone ratio of 1:1. Further, the β-carotene extract stored in dark bottle for one month exhibited improved stability with superior β-carotene content, i.e., 92.18 µg/ml in contrast to the extract stored within a translucent container (20.12 µg/ml). Minh et al. (2015) demonstrated successful β-carotene extraction from sweet potato using n-hexane as solvent at 80°C in four hours.

3.10 Sweet Potato Ice Cream

Ice cream is one of the most preferred desserts worldwide, where it can be served in various forms; in cups, cones, topped with chocolates etc. In the last two decades many new formulations have been included in the preparation of ice cream. Different varieties of fruits

Table 3.1. Unfermented food products from sweet potato.

Products	Preparation technology	Reference
Sweet potato flour	Peeling, slicing, drying and grinding.	Owori and Hagenimana 1999
Bakery products and biscuits	Blending of sweet potato flour with wheat flour along with necessary ingredients of appropriate proportions for dough making and baking.	Mu et al. (2017) Perez et al. (2017) Mais (2008)
Starch	Extraction by wet separation technique of the ground crop.	Mu et al. 2017
Slices and chips	Slicing of the roots upto 1 cm in thickness and further application of different drying processes such as cross flow, greenhouse solar, and open air-sun.	Nicanuru et al. 2015 Bechoff et al. 2009
Noodles and pasta	Slurry mixing, stirring, gelatinization of starch and processing machine by screw extrusion.	Mu et al. 2017
Puree	Washing, peeling of roots, trimming and cutting is carried out. Finally the puree is produced by steam cooking.	Fasina et al. 2003 Wanjuu et al. 2018
Baby food	Grinding and mixing of sweet potato flour with other protein rich substrates. Precooking and drying. Mixing with appropriate quantities of vitamins and minerals as per the norms.	Nandutu and Howell (2009) Laryea et al. 2018
Anthocyanin	Extraction of pigments of purple fleshed sweet potato using polar solvents, such as ethanolic and methanolic solvents and acidified electrolyzed water.	Kim et al. (1996) Lu et al. 2010
Ice cream	Incorporation of purple sweet potato in a particular proportion to the ingredients for ice cream preparation.	Kale (2017) Weenuttranon (2018)

and other ingredients have been used to develop unconventional ice cream flavours rather than just the traditional vanilla and strawberry ice creams. Purple sweet potato has also been used as an ingredient in the preparation of ice cream. Recently, a study was conducted by Weenuttranon (2018), which demonstrated the application of purple sweet potato for an improved organoleptic response. This was observed by studying the sensory characteristics of the ice cream that by adding of 10%, 20% and 30% of purple sweet potato in the recipe. It was found that the supplementation of 30% of the purple sweet potato was the most preferred recipe (appearance and the average color range was 7.00 and 7.16, respectively) in the production of ice cream in contrast to the ice cream prepared by 10% (appearance, 6.42; average color range, 6.12) and 20% (appearance, 6.56; average color range, 6.36) purple sweet potato. Sweet potato waffle ice cream cones are prepared by whisking the boiled sweet potatoes with milk, eggs, honey, vanilla and butter, further the ingredients are folded into the flour premixed with baking powder and poured on a hot griddle (https://www.chemours. com/Teflon/en_US/assets/downloads/pdf/201407_Sweet_Potato_ Waffle.pdf). In another piece of research (Kale 2017) it was reported that the use of fat in ice cream can be partially replaced by the specially modified starch of sweet potato (upto 20%). The starch of the sweet potato is treated with an enzyme and hydrothermal process for a preferable quality to the ice cream. The ice cream produced by using sweet potato modified starch was feasible in a technoeconomical perspective and the cost of production was 104.41–108.89 per kg. The unfermented sweet potato products have been presented in Table 3.1.

Review on Analysis of Wine and Beer

4.1 Biochemical Studies of Alcoholic Beverages

Biochemical study is a very important aspect in the analyses of alcoholic beverages. The study includes various parameters like TSS (total soluble solids), starch, total sugar, pH, specific gravity, titratable acidity, tannin, anthocyanin, ethanol, phenol, lactic acid, ethanol and DPPH activity. Various studies have been conducted in preparation of wines based on the above parameters. In the preparation of cashew (*Anacardium occidentale* L.) wine, the must and wine had TSS (°Brix) of 17.0 and 2.0, reducing sugar (g/100 ml) of 6.44 and 0.9, titratable acidity (g tartaric acid/100 ml) of 0.24 and 1.21, pH of 4.63 and 2.92, phenol (g/100 ml) of 0.13 and 0.12, tannin (mg/100 ml) of 2.2 and 1.9, respectively. The cashew wine contained 5(%) ethanol (Mohanty et al. 2006). Similarly, biochemical analyses have been conducted to study the analytical variables in the must and wine of jamun fruits (*Syzgium cumini* L.). The must and wine had TSS (°Brix) of 16.5 and 2.8, reducing sugar (g/100 ml) of 6.48 and 0.49, titratable acidity (g tartaric acid/100 ml) of 0.51 and 1.11, pH of 4.50 and 3.30, phenol (g/100 ml) of 0.23 and 0.22, tannin (mg/100 ml) of 1.70 and 1.40, anthocyanin (mg/100 ml) of 85 and 60 and ethanol of 0 and 6% (v/v), respectively (Chowdhury and Ray 2007). Biochemical studies have also been conducted in other tropical wine preparations like litchi wine (Kumar et al.

2008), mango wine (Reddy and Reddy 2005), gabiroba wine (Duarte et al. 2009), etc. Similar studies have been conducted in the preparation of herbal wine. Wine prepared from infused tea leaves of two different types of clones, LT 318 and LT 143 had a pH of 3.33 and 3.60; specific gravity of 1.0299 and 1.0281; total soluble solids of 5.70% and 5.60% and ethanol of 6.0% (v/v) and 7.2% (v/v), respectively (Aroyeun et al. 2005). Like wine, wort and beer are also subjected to biochemical analysis. Sweet sorghum beer was subjected to biochemical analysis based on the parameters like tannin, pH, ethanol, reducing sugar, etc. The beer had a pH of 4.16, tannin of 5.33 mg/100 ml, alcohol of 4.80% and residual reducing sugar of 4.10 mg/g (Mesta 2005).

4.2 Fourier Transform Infrared Spectroscopy Studies in Alcoholic Beverages

Fourier transform infrared (FTIR) studies are conducted to identify either organic or inorganic chemicals from an unknown mixture. Infrared (IR) spectroscopy has been considered as a reliable and non-invasive modern tool for the determination of the quality of alcoholic beverages such as wine, beer and spirits. Further, the special advantage involved with infrared studies is the capacity to quantify the concentration of ethanol in alcoholic beverages as well as it is also useful in identification of the unwanted and risky contaminants such as methanol. The infrared data in combination with statistical analysis such as principal component analysis and partial least square provides a faster method to determine the quantity of ethanol and contaminants as well (Ramasami et al. 2005). The application of IR is cost effective; hence it is applied to study the composition of alcoholic beverages, compositional change in fermentation and organoleptic profiling (Gishen et al. 2006). Only two minutes are required for the IR analysis of alcoholic beverages. For distilled alcoholic beverages there is no requirement for the preparation of the sample whereas in carbonated alcoholic beverages such as beer and champagne, degassing has to be done prior to the analysis (Lachenmeier 2007). Recently, an interesting study was conducted to estimate the ethanol contents of popular indigenous fermented (alcoholic) beverages of Ethiopia using mid infrared spectroscopy. Mid infrared (1200–850/cm) coupled with partial least square was applied to determine the concentration of ethanol.

Amongst the studied beverages, i.e., *Tej* (honey wine), *Tella* (a malt beverage like beer), *Korefe, Keribo, Birz, Netch Tella, Filter Tella* and *Borde* the highest and the lowest ethanol content were observed in *Tej* (9.1%, v/v) and *Keribo* (0.77%, v/v), respectively (Debebe et al. 2017).

4.3 Sensory Evaluation of Alcoholic Beverages

To determine the acceptability of a newly developed product, sensory evaluation is conducted. Various studies have been conducted to ensure the quality of the product as well as its comparison with other similar commercial products. In the production of litchi wine, sensory evaluation has been conducted to determine the consumer acceptability. Sensory attributes were estimated through a 5–point Hedonic scale (1 = dislike extremely and 5 = like extremely) by 16 panelists comprising of 8 men and women each, aged 20–35 years. Principal component analysis (PCA) reduced the five original sensory parameters (taste, aroma, flavour, colour/appearance and aftertaste) to three independent principal components that are responsible for 79.3% of the variation of the wine (Kumar et al. 2008).

Considering the major parameters the sensory panelists rated the reference grape wine to be superior to that of the jamun wine. The high tannin content of the jamun wine is responsible for the astringency of jamun wine and the flavour scoring fell in between 2 to 3 (like moderately to like much) (Chowdhury and Ray 2007). Wine produced from infused tea leaves, LT 318 was rated the best among the samples produced from LT 143 and another processed tea (Unilever Plc. Nigeria) based on taste, aroma, attributes, colour and overall acceptability (Aroyeun et al. 2005).

4.4 Statistical Analysis of Fermented Food Products

Principal component analysis (PCA) is a technique adapted expediently for the assessment in the analysis of food products (Arvanitoyannis and Tzouros 2005). In PCA, data on huge number of variables are reduced into small components without affecting the majority of the data (Hair et al. 1998). The important characteristic of the process is diminution of the dimensionality in a set of variables by constructing an uncorrelated linear combination of them. The combinations are calculated in such a manner that the first component represents for the major part of variance, that is the major axis of the points in the *p*-dimensional space (Tzouros and Arvanitoyannis 2001).

Principal component analysis (PCA) was applied to litchi wine which reduced the eight original analyses (proximate) variables [TSS (°Brix), reducing sugar (g/100 ml), titratable acidity (g tartaric acid/100 ml), pH, phenol (g/100 ml), tannin (mg/100 ml), lactic acid (mg/100 ml), and ethanol (%)] into two independent components (factors, PC1 and PC2), that showed 100% of the variation in litchi wine (Kumar et al. 2008). Kohajdova et al. 2007 conducted a study on lactic acid fermented cucumber juice and the biochemical variables were condensed to two principal components, i.e., PC 1 and PC 2. PC 1 that accounted for 71.6% of the variations and PC 2 showed 16.5% of total variations (Kohajdova et al. 2007).

Pearson's correlation is applied to the analytical variables of food and beverage samples to determine the degree of relation between different parameters.

In case of litchi wine the major correlations found were TSS–TA (–0.866), TA–pH (0.925), TA–phenol (0.866), phenol–tannin (0.786), etc. (Kumar et al. 2008). Similarly strong correlations were found between total acidity and pH (–0.98), total acidity and content of organic acids (0.91) and between pH and content of organic acids (–0.88) in case of fermented cabbage juice (Karovikova et al. 2001).

Analysis of Variance (ANOVA) is one of the common statistical analyses used for scientific experiments. ANOVA is applied for comparisons of the means of multiple samples either from the same group/population or different group/population (O'Mahony 1986). In addition to this, a *post-hoc* analysis is conducted to understand the details of the variations. ANOVA has been widely used in analysis of food products. It is applied to interpret the organoleptic differences and comparative studies. Maeztu et al. (2001) have conducted descriptive analysis of ANOVA for organoleptic characteristics of coffees. Similarly organoleptic profiling of cheeses (Murray and Delahunty 2000) and fluid milk (Claassen and Lawless 1992) have also been conducted. In another study Phillips et al. (1995) have applied ANOVA to study the variations among milk with wide-ranging percentage of fat.

4.5 Technoeconomical Feasibility and Cost Economics

Technoeconomical feasibility and cost economics is studied when a new process or technology is developed. The study enables the innovator or the investor to understand the investment, product output, cash flows as well as the economic viability of any new industry/process. Few studies have been conducted on the technoeconomical feasibility and cost economics of food and beverage production. Zhang and Rosentrater (2015) demonstrated the technoeconomic analysis and lifecycle assessment of small, medium and large scale wine making technology. The findings of life cycle assessment revealed that manufacture of the bottle as well as wine making had the highest impact on the environment where as in technoeconomical assessment of cost and profit, all the scales were exponentially related. In another study, an evaluation of engineering economics was done for the commercial production of cookies by using dried brewers' spent grain, groundnut cake and sorghum flour as substrate. The study was based on assumptions [cash flow was considered to be consistent over the plant life, i.e., one decade with no salvage value, manufacture of the cookies was based on constant mass flow rate of 90 packets/min and effects of uncertainties on cookie production were calculated by varying the operation days (330, 300 and 250 days)]. It was found that the capital cost of the industry was US$ 1.39×10^6, and the annual production cost was US$ 10.08×10^6/year. Further, the return on investment (ROI) was 63% whereas the gross margin and internal rate of return (IRR) of the production unit were 23% and 77.52%, respectively (Olawoye et al. 2017).

5

Materials and Methods for the Preparation of Sweet Potato Alcoholic Beverages

◇◇

5.1 Purple Sweet Potato Roots

Roots of the purple-fleshed sweet potato were employed for preparation of three unique beverages (alcoholic fermented), i.e., (1) anthocyanin rich wine (red wine), (2) herbal purple sweet potato wine, and (3) anthocyanin rich beer.

Healthy roots were harvested and the fresh roots were immediately brought to the laboratory to conduct formulations of the aforesaid wine and beer during the winter season (October–January) of the year 2009–2011. Purple sweet potato roots were collected from three sections (4 m × 4 m) of a bigger plot (40 m × 40 m) and were mixed up properly for further use.

The biochemical constituents of purple sweet potato were estimated as follows and has been presented in Chapter 7.

5.2 Barley, Malt and Hop

Barley (*Hordeum vulgare*), a major ingredient in the preparation of beer, was purchased from a local market in Bhubaneswar, Odisha. In the process of malting, barley was immersed in water; the excess water was drained out to kick off the germination of the seedlings from the seeds. During the germination of the seeds, enzymes such as amylases and proteases are activated, that are responsible for the breakdown of starchy and protein sources into simple sugars and amino acids, respectively which the emerging plants would use. The malted barley was further dried and crushed by using a mixer-cum-grinder to prepare grist. Bio-chemical analyses were carried out to determine the composition of both the barley and malt.

Hops in the form of pellets were obtained from Arun and Co, Mumbai as gift.

5.3 Yeast

The common yeast strain, *Saccharomyces cerevisiae*, maintained on Potato Dextrose Agar (PDA) slants at 4°C, is frequently applied in preparation of alcoholic beverages in our laboratory (Mohanty et al. 2006, Chowdhury and Ray 2007, Kumar et al. 2008). This yeast has been used to prepare all the three novel fermented alcoholic products (anthocyanin rich wine, herbal purple sweet potato wine and anthocyanin rich beer) described in this book.

5.4 Enzymes for Saccharification

Three different enzymes were procured to be used in the preparation of wine and beer. Termamyl® and Dextrozyme G.A were applied in the saccharification of the purple sweet potato roots during the preparation of wine. Termamyl® functions at a temperature of >90°C and the appropriate working temperature for the Dextrozyme G.A is >45°C. Promalt is a mixture of amylase and protease which is used in preparation of wort in the beer preparation.

5.5 Anthocyanin Rich Purple Sweet Potato Wine (Red Wine)

5.5.1 Fermentation process

Sweet potato is rich in starch content; hence saccharification (breakdown of starch into fermentable sugars) by enzymatic treatment is an essential step for preparation of purple sweet potato wine.

5.5.1.1 Saccharification

Healthy fresh purple sweet potato roots were obtained and properly cleaned under running tap water to rinse out the soil adhered to it and then wiped dry by using a dirt free fabric. One kg of the roots were peeled, mashed and uniformly mixed with one litre of water in a mixer-cum-grinder. The purple sweet potato slurry was applied with a thermostable α-amylase, Termamyl® at a concentration of 0.2% and further incubated at 90°C for 1hr for liquefaction. After the liquefaction, the temperature of the mash was brought down to below 45°C by cooling and Dextrozyme G.A of 1% concentration was incorporated through proper mixing. The mixture was incubated for 48 hr at 45°C in a BOD incubator for breakdown of starch.

The saccharified slurry was allowed to cool to normal temperature (28±2°C) and the juice, known as must, was filtered out through a clean and sterilized cotton fabric. The filtered must was first subjected to biochemical analysis to check the sugar content. Later additional cane sugar was added to increase the total soluble solids to 20°Brix from the initial content of 16°Brix. Sodium metabisulfite was put in to the must at a dose of 100 µg/ml in order to prevent the growth of unwanted microbes such as *Acetobacter* sp., moulds, etc. (Amerine and Ough 1980). The ameliorated must was inoculated with a 24 hr old starter culture of wine yeast (*S. cerevisae*) at a rate of 2% along with 0.1% concentration of $(NH_4)_2SO_4$ as a source of nitrogen for proper growth of the wine yeast. Fermentation of the wine was conducted in an ambient temperature, i.e., 28±2°C for 120 hrs or five days. The experiment was carried out with three replications.

5.5.1.2 Starter culture

Hundred grams of healthy and fresh grapes were washed thoroughly, mashed in a mixer cum grinder and the juice was extracted through a sterilized clean cotton cloth. The juice was allowed to boil for 10–15 min. Equivalent volume of sterile water was blended with the extracted juice and cooled to normal temperature. Further, the juice was inoculated with *S. cerevisiae* (laboratory stock culture) under necessary aseptic conditions in a laminar air flow and incubated at 30°C for 24 hr for use as starter culture.

5.5.2 Racking and bottling

After five days of fermentation, racking was conducted. Racking is a process of decantation of the clear top wine from the entire fluid while leaving the lees in the bottom. Racking was conducted repeatedly for three times in every 20 days. Bentonite was added to the wine before the final racking was carried out after the complete settlement of the last residues of the wine. Sodium metabisulfite (100 µg/ml) was again incorporated to act against unwanted microbes for long term storage. The sterile and clean glass bottles were filled completely and corked. Bee wax was used as sealing agent. The flowchart for preparation of purple sweet potato wine is presented in Fig. 5.1.

5.5.3 Preparation of samples for FTIR analysis

FTIR study was conducted for purple sweet potato wine and a reference wine [grape wine (Nashik Vintners Pvt. Ltd, Maharashtra, India)] to determine the differences in compositions. Two ml of each of the wine samples were put on an oval glass piece with diameter of 1×1 cm^{-1} and air dried overnight. The spectra of the dried samples were recorded by using a Perkin-Elmer 2000 spectrometer (Perkin-Elmer, Norwalk, CT) at 4/cm resolution with a narrow band liquid N_2-cooled MCT (Mercury Cadmium Telluride) detector.

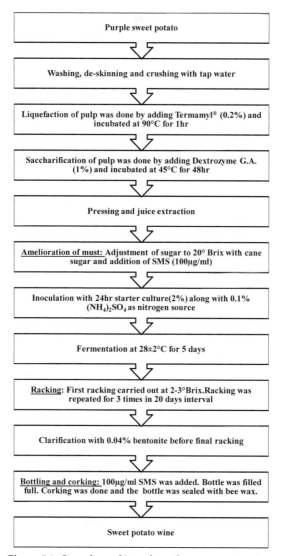

Figure 5.1. Steps for making of purple sweet potato wine.

5.5.4 Sensory/organoleptic assay

The attributes considered for sensory evaluation were clarity, colour, viscosity, odour, flavour, spritz and finish. The assessments of the parameters were carried out in a 9-point Hedonic scale (1 = dislike extremely and 9 = like extremely) by 16 panelists who were familiar to alcoholic beverages, especially wine as conducted by Mohanty et al. (2006). The panelists were trained about the different parameters and the questionnaires were discussed before conducting the organoleptic assay (Kilcast and Subramanian 2000). A blind study was conducted to get the impartial results. The samples were coded randomly with a three digit code. Fifty ml of the purple sweet potato wine and a commercial grape wine were served to the panelists in clean transparent wine glasses. Water was provided to the panelists to rinse their mouth in between the evaluation of different parameters of the wines. The same experiment was replicated the next day and the mean of the data were considered as the final scores.

5.5.5 Statistical methods

Paired 't' test and PCA (principal component analysis) were applied to the biochemical variables using statistical software SPSS (SPSS Software for Windows release 17.0 SPSS Inc., Chicago, IL, USA). PCA is applied to group the biochemical variables into smaller components for better analysis of their relationships whereas paired 't' test was conducted to determine the mean differences in the biochemical variables of both the must sample and wine fermented using *S. cerevisiae*. The organoleptic scores of the various parameters (clarity, colour, viscosity, odour, flavour, spritz and finish) of both purple sweet potato wine and reference wine were subjected to independent 't' test for determination of the mean differences of the aforementioned parameters.

5.6 Herbal/Medicinal Wine from Purple Sweet Potato

5.6.1 Medicinal herbs and plant parts

Different healthy plant parts of medicinal importance were collected from the local garden during the month of November 2009 (day temperature, $28 \pm 2°C$ and night temperature, $20 \pm 2°C$). They were rinsed properly and sundried and further crushed. The details of the specific herbal parts are presented in Table 5.1.

5.6.2 Fermentation process

The procedure followed for the fermentation of herbal sweet potato wine is similar to that of the red wine produced from sweet potato (Section 5.5.1).

5.6.2.1 Saccharification

Saccharification of the purple sweet potato roots were conducted by using enzymes, thermostable α-amylase Termamyl and Dextrozyme G.A in a similar manner described in the Section 5.5.1.1.

5.6.2.2 Preparation of starter culture

A 24 hr old starter culture of *S. cerevisiae* was prepared in the same process as described in the preparation of anthocyanin rich purple sweet potato wine.

5.6.3 Racking and bottling

After five days of fermentation, racking was conducted as described above for purple sweet potato wine (Section 5.5.2). Sodium metabisulfite (100 µg/ml) was again added as a preservative to inhibit unwanted microbes for long term storage. The sterile and clean glass bottles were filled completely and corked. Bee wax was used as sealing agent. The flowchart for preparation of purple sweet potato wine is presented in Fig. 5.2.

5.6.4 Preparation of samples for FTIR analysis

FTIR analysis was conducted for both purple sweet potato herbal wine and a reference grape wine (Nashik Vintners Pvt. Ltd, Maharashtra,

Table 5.1. Medicinal herbs/plant parts and quantity used in the preparation of herbal purple sweet potato wine (Jain 1983).

Common english name	Botanical name	Family	Part of the plant used	Quantity (g/kg sweet potato)	Medicinal properties
Ink nut	*Terminalia chebula*	Combretaceae	Fruit	5	anti-asthmatic, anti-dysenteric, anti-paralytic in piles
Belliric myrobalan	*Terminalia belerica*	Combretaceae	Fruit	5	anthelmintic, antiseptic, astringent, expectorant, laxative, lithotriptic, rejuvenative
Indian gooseberry	*Emblica officinalis*	Phyllantheae	Fruit	5	diuretic, laxative, carminative, stomachic, astringent, antidiarrhoeal, antihaemorrhagic and antianaemic
Ginger	*Zingiber officinale*	Zingiberaceae	Rhizome	3	anti-dyspepsia and anti-cancer
Bael	*Aegle marmelos*	Rutaceae	Fruit	5	anti-chronic diarrhea, anti-dysentery, anti-peptic ulcers and laxative
Bishop's weed	*Carum copticum*	Apiaceae	Fruit	3	antispasmodic, antiseptic
Holy basil	*Ocimum sanctum* L.	Lamiaceae	Leaves	5	anti-bronchial, anti-asthmatic, anti-malarial, anti-dysenteric, anti-arthritis
Night jasmine	*Nyctanthes arbortristis*	Oleaceae	Leaves	2	anti-sciatica and anti-arthritic, antipyretic and ulcerogenic activity
Malabar nut	*Justcia adhatoda*	Acanthaceae	Leaves	2	antispasmodic, bronchodilator and mucolytic agent in asthma
Five leaved chaste tree	*Vitex negundo*	Lamiaceae	Leaves	2	anti-inflamatory, antipyretic, tranquilizer, bronchial smooth muscle relaxant, anti-arthritic, anti-helminthic, and vermifuge

68

Garlic	*Allium sativum*	Amaryllidaceae	Fruit	3	hypolipidemic, antithrombotic, antioxidant, anti-cancerous and anti-microbial effects
Cardamom	*Elleteria cardamomum*	Zingiberaceae	Fruit	2	anti-asthma, anesthetic in burning sensation, anti-cold and cough, protects from diseases of bladder and kidney
Cinnamon	*Cinnamonum verum*	Lauraceae	Fruit	2	anti-flatulence, anti-piles, anti-amenorrheal, anti-diarrhoeal
Indian aloe	*Aloe vera*	Xanthorrhoeaceae	Leaves	2	anesthetize tissues, acts against bacterial, fungal and viral growth, anti-inflammatory, dilate capillaries, anti-cancer and enhance blood flow
Belladonna	*Atropa belladonna*	Solanaceae	Root	2	promoter for wound healing
Asparagus	*Asparagus adscendens*	Liliaceae	Root	2	anti-spermatorrhoea, anti-chronic leucorrhoea, anti-diarrhoea, anti-dysenteric symptom
Round leaf buchu	*Barosma betulina*	Rutaceae	Leaves	2	stimulates kidney, cleanses blood, antiseptic, acts as tonic and promotes sweating
Neem	*Azadirachta indica*	Meliaceae	Bark	0.5	analgesic, alternative, curative of fever

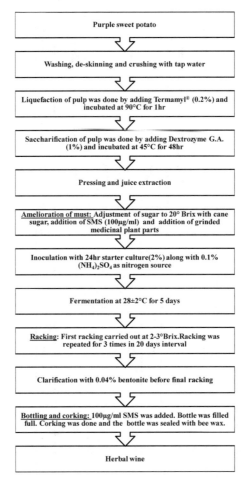

Figure 5.2. Steps for making of herbal purple sweet potato wine.

India) for the assessment of the difference in compositions. The method used for the analysis was the same as that of the process used in the analysis of the purple sweet potato wine.

5.6.5 Statistical methods

The biochemical parameters of purple sweet potato herbal wine were subjected to PCA (principal component analysis) to group the related variables into smaller components for understanding the relationship among them using the same SPSS software that was used for purple sweet potato wine.

5.7 Anthocyanin Rich Beer from Purple Sweet Potato

5.7.1 *Purple sweet potato flakes*

Freshly harvested purple sweet potato roots were washed, wiped and de-skinned and the peeled roots were chopped into smaller pieces. The chopped pieces were then sundried in an open clean space covered with net with 0.6 mm mesh size to protect it from insects and pests. The sun drying continued until the moisture content reduced to 6–8% and roots were made into flakes. The purple sweet potato flakes were stored in an air tight container for applications.

5.7.2 *Barley grains and hops*

Healthy good quality barley (*Hordeum vulgare*) grains were bought from the local market of Bhubaneswar, Orissa. Hops were obtained in form of pellets.

5.7.2.1 *Malt preparation*

The malt was prepared in the standard procedure of malting. It is well adapted by breweries and other food industries. Firstly, the grains were properly washed to remove the impurities and unwanted substances. Further, it was soaked in water for 24 hrs, and the water was changed every 6 hrs of thee soaking period. This process is called steeping, which takes place in aerobic conditions. By the time soaking and steeping is complete, generally 45% water is absorbed by the grains. The grains were then placed for germination. Germination for growth of new barley seedlings were allowed until the acrospires (cotyledon/ shoot) grew to about 3/4th size of the grain size. The germinated grains are known as green malt. The green malt is processed to malt by kilning. Kilning was conducted by heating the green malt in an oven at 80–100°C for 24 hours to produce malt. The malt was brought out of oven and left to reach room temperature (28±2°C). Finally, it was grinded to obtain grist, which was used in our study for the preparation of purple sweet potato beer.

5.7.3 Fermentation process

5.7.3.1 Preparation of wort

Wort was prepared by taking three different proportions of purple sweet potato flakes and grist obtained from malted barley [30% purple sweet potato flakes + 70% grist, 50% purple sweet potato flakes + 50% grist, and 100% purple sweet potato flakes + 0% grist] and a control sample of 100% grist was also prepared. 300 g of the substrates of all the appropriate proportions were used for preparation of all the three types of beer of 2 l each (triplicates) along with control beer (prepared from 100% grist). Each substrate combination was properly blended with 1 litre water in different vessels prior to saccharification.

5.7.3.2 Saccharification

Promalt [Novozyme, Denmark], a mixture of amylase and protease was incorporated to the slurry (1 l water and 300 g substrate) at a concentration of 0.1 g/l and mixed. The enzyme mixed slurry was then incubated at 42°C, 50°C, 62°C, 70°C in sequence for 20 minutes each for activation of the enzyme promalt. During the saccharification, the starch content of grist and purple sweet potato are broken down to fermentable sugars.

5.7.3.3 Boiling

The saccharified slurry was boiled together with hop pellets (0.075 g/l) for 90 min in a stainless steel kettle. Boiling water was added to the boiling wort to maintain the volume at 2 l. At that time a dose of 40 g/l of boiling dextrose was added. After boiling, the specific gravity was measured and adjusted to 1.040 by addition of sterilized dextrose. Further, the wort was cooled down to ambient temperature

($28 \pm 2°C$) and sieved in a sterilized clean cotton cloth to obtain the final wort for fermentation.

5.7.3.4 *Preparation of starter culture*

A 24hr old starter culture of *S. cerevisiae* was prepared by using MYGP (malt extract 3.0 g/l; yeast extract, 3.0 g/l; glucose, 10.0 g/l; peptone, 5.0 g/l) broth medium (incubation temperature, 20°C). 10% of the starter culture was inoculated to each of the 12 samples (triplicates of wort prepared in different combinations of purple sweet potato flakes and grists).

Fermentation was conducted for five days at 20°C (maintained in a BOD incubator). After measuring the specific gravity, fermentation was further allowed until the specific gravity of the medium fell to 1.010.

5.7.4 *Finishing and bottling*

After the fermentation was over, the beer was stored at 4°C so that the residual suspended particles could settle. The upper portion of the fluid was carefully poured out and filtered through a sterilized cheese cloth without disturbing the bottom residues. Beer was filled in to dark bottles. Priming sugar 3 g/l was added to the beer filled bottle for carbonation and foaming. Finally, the dark bottles filled with beer were sealed with a crown using a hand packaging machine (Bajaj Machines, New Delhi). The purple sweet potato beer making flowchart has been presented in Fig. 5.3.

5.7.5 *FTIR analysis*

Infrared spectroscopic study was conducted with different beer samples developed during our study. The FTIR graph of each of the purple sweet potato beer samples have been represented with reference to the control sample (100% grist). The method used for the analysis

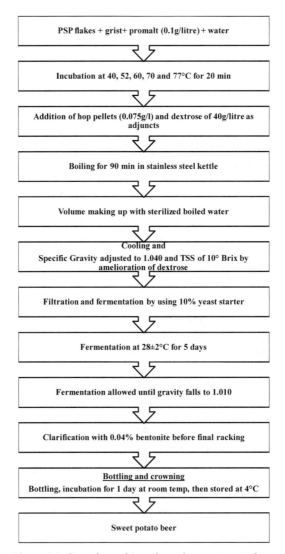

Figure 5.3. Steps for making of purple sweet potato beer.

was the same as that of the process used in the analysis of the purple sweet potato wine and herbal wine.

5.7.6 Sensory evaluation of beer

A 30 member trained panel was selected for the assessment of the organoleptic properties of purple sweet potato beer samples. Like the purple sweet potato wine, the members of the panel were trained about the purple sweet potato beer, definition of sensory/organoleptic attributes and questionnaires. The different sensory attributes considered for the study were colour, clarity, foam and flavour on a nine point hedonic scale (1—Dislike extremely, 9—Like extremely). During the assessment, each panelist was provided with 100 ml of coded beer samples in polythene cups.

5.7.7 Statistical analysis

ANOVA was applied to the purple sweet potato wort and beer samples to determine the mean differences among the beer samples prepared from different proportions of purple sweet potato flakes and grist. When significant differences in ANOVA ($p < 0.05$) were detected, the Tukey's HSD multiple comparison tests were useful to evaluate the factor level differences. Further the biochemical data were subjected to principal component analysis (PCA) to extract the factors based on the correlational relationships among the variables. The factors were rotated using Varimax Rotation. The factors that exceeded the eigen value of 1 were retained for analysis. In the rotated component matrix, when variable loading on a particular component exceeded 0.72, it was chosen to represent the factor (Field 2000). All the analyses were performed using SPSS (SPSS Software for Windows release 17.0, SPSS Inc., USA).

Materials and Methods for Biochemical and Technoeconomical Feasibility (Large Scale) Studies

◇◇◇

6.1 Biochemical Analysis

Various biochemical analyses were carried out for the purple sweet potato roots, barley, malt, must, wine, wort and beer. Experiments were carried out in three replicates for each sample. Data from the analysis were represented as mean of the results of the triplicates.

6.1.1 Biochemical analysis of purple sweet potato roots

The freshly harvested purple sweet potato roots were de-skinned and the analyses for different biochemical parameters of the roots were carried out as described below.

6.1.1.1 Estimation of moisture

Ten gram of the sample was placed on glass petri–dishes and was subjected to drying in an oven for 8–12 hr at $100 \pm 2°C$ till an unvarying weight was achieved. The sample was further cooled to room temperature. The loss in weight was articulated as % moisture and computed by the formula:

$$\text{Moisture (\%)} = \frac{(b-a)-(c-a)}{(b-a)} \times 100$$

Where,

a = empty weight (g) of glass petri–dishes

b = weight (g) of petri–dishes + samples before oven drying

c = weight (g) of petri–dishes + samples after oven drying

6.1.1.2 Estimation of crude protein

6.1.1.2.1 Reagents

a. Digestion mixture

b. N/7 H_2SO_4

6.1.1.2.2. Method

The crude protein content of the purple sweet potato roots were estimated by the Kjeldahl method (AOAC 1984). A 500 mg sample was put into a Kjeldahl flask with 10 ml of concentrated sulfuric acid trailed by adding up of the digestion mixture described above for digestion at 42°C for 2 hr. After the digestion was complete, the sample was cooled and distilled with 10–20 ml of 40% NaOH. Subsequently, titration of the distillate was carried out against standard N/7 H_2SO_4 till a red colour emerged and the crude protein content was estimated by the formula:

$$\text{Crude protein (\%)} = \frac{100 \times y \times (b-b_1) \times 0.02 \times 6.25}{x \times w}$$

Where,

y = volume (ml) made out of digested sample

b = volume (ml) of N/7 H_2SO_4 consumed

b_1 = volume (ml) of N/7 H_2SO_4 consumed for blank

w = weight of dried sample

6.25 = factor for converting nitrogen into protein

6.1.1.3 Estimation of total sugar and starch

6.1.1.3.1 Alcoholic extract preparation

As purple sweet potato roots contain a high amount of starch and sugars, only 10 mg of the purple sweet potato root cube was homogenized with 5 ml of ethanol (80%) and transferred to test tubes. In the case of liquid samples, diluted samples (1000 times) was meticulously blended with 5 ml of 80% ethanol and further shifted to test tubes. The test tubes were subjected to a hot water bath at 80–85°C for 10 min and cooled. Centrifugation is carried out at a rate of 4000 rpm for 15 min. The supernatant obtained during the centrifugation was stored and the same process was repeated thrice. All the supernatant samples collected were placed in a single beaker and kept under the fan at room temperature (30±2°C) for the evaporation of ethyl alcohol.

Then 2 ml of distilled water was added to the residues followed by 2 ml of 2N HCL and then the tubes were put in a boiling water bath for about 20 min with intermittent stirring. Absence of blue colour with N/10 HCL iodine solution reaction indicated the completion of the hydrolysis. The tubes were cooled and centrifuged at 4000 rpm for 10 min, finally the supernatants were made to the volume of 10 ml by the addition of distilled water. These extracts were used for estimation of starch, as a glucose equivalent.

6.1.1.3.2 Total sugar

6.1.1.3.3 Reagent

a. 0.2 % Anthrone reagent

6.1.1.3.4 Method

Two ml anthrone reagent was taken in a test tube and chilled in an ice bath and subsequently 1 ml of alcoholic extract was carefully laid over the reagent. The test tube was then strongly shaken inside in the ice–bath. The tube was brought to room temperature followed by boiling in the water bath for around 10 min. The blank sample without the alcoholic extract was prepared in similar manner. After cooling, the absorbency of the samples was recorded at 625 nm. Total sugar content was evaluated following a standard curve plotted with 0–50 µg/ml of D–glucose per assay (Mahadevan and Sridhar 1998).

6.1.1.3.5 Starch

Starch content was evaluated by multiplying the total sugar content by a factor 0.9 (Mahadevan and Sridhar 1998).

6.1.1.4 Estimation of dietary fibre

6.1.1.4.1 Reagents

   a. NaH_2SO_4 (2.04N)
   b. NaOH (2.5N)

Two grams of moisture and a fat free sample was taken in a 200 ml spoutless beaker and subsequently 2.04 N NaH_2SO_4 (25 ml) was added. Volume was made up to 200 ml. Further, the beaker was placed on a warmed hot plate and the content was refluxed for half an hour. Filtration of the content was done using a Buchner funnel and the left out residues were rinsed with hot water to make it acid free. Following this, the residues were further extracted with 25 ml of 2.50 N NaOH and the residues left were taken in silica basins using a spatula. The residues were oven dried at $100 \pm 2°C$. Then these residues were cooled, weighed along with silica basins and decarbonized on heater at 550–600°C for 1–2 hr. The ash left on the silica basins was weighed after cooling in the desiccator to find out the dietary fibre contents using the formula:

$$\text{Dietary fibre (\%)} = \frac{(a - b)}{w} \times 100$$

Where,

a = weight (g) of silica basins + residues left after acid and alkali digestion

b = weight (g) of silica basins + ash

w = weight (g) of oven dried samples

6.1.1.5 Estimation of ash

A muffle furnace was used to determine the ash content in any sample. Samples in powder form (1 g) were acquired in pre–weighed silica basins and again combined weights were noted. The sample was decarbonized on a heater and shifted along with basin to the muffle furnace following ignition at 550–660°C for 2–3 hr. Later, the silica basins were cooled in a desiccator to get the ash contents of samples by using the following formula:

$$\text{Total ash (\%)} = \frac{(c-a)}{(b-a)} \times 100$$

Where,

a = empty wt (g) of silica basins

b = weight (g) of basins + samples

c = weight (g) of basins + ash

6.1.1.6 Quantification of elements (minerals)

6.1.1.6.1 Sample preparation

Roots were thoroughly washed and rinsed in running tap water and then with distilled water to discard any possible surface contamination and dried at 55–60°C in an electric oven. Homogenization of the dried samples was done using a pestle mortar to make it into a powder form.

6.1.1.6.2 Reagents

a. HNO_3

b. H_2SO_4

c. $HClO_4$

6.1.1.6.3 Method

Powered sample of purple sweet potato weighing 0.25 g was digested in 6.5 ml of acid solution (HNO_3 : H_2SO_4 : $HClO_4$ in ratio of 5:1:0.5). The mixture was heated until white fumes had appeared. Dilution of the clear solution was done using distilled water up to 50 ml followed by filtration with Whatman filter paper No. 1. The standard working solutions of the selected elements (Ca, P, Mg, Na, K, S, Fe, Cu, Zn, Mn, Al and B) were prepared to make the standard calibration curve. Absorption of the sample solution was used against the calibration curve to determine the concentration of the particular element in the sample. Atomic Absorption Spectrophotometer (Perkin–Elmer, AAnalyst–600/800, Norwalk, CT) was used to determine the 12 elemental concentration in the sample. Air–acetylene gas was used for conducting all of the experiments.

6.1.1.7 Estimation of anthocyanin

6.1.1.7.1 Reagents

a. HCl in aqueous methanol – 0.5 N HCl in 80–85% methanol
b. Hydrogen peroxide reagent – 1 ml 30% H_2O_2 + 9 ml methanolic HCl

6.1.1.7.2 Sample preparation

Peeled root samples of 0.5 g samples were extracted with 80% ethyl alcohol using a mortar and pestle and the volume was made up to 25 ml.

6.1.1.7.3 Method

Sample (2 ml) was taken in a test tube to which 3 ml of HCL in aqueous methanol was added followed by 1 ml of hydrogen peroxide reagent. A blank/control solution was prepared with methanolic HCL only. Then these test tubes were kept in the dark for 15 min and absorbency of these samples was taken at 525 nm (Mahadevan and Sridhar 1998). The anthocyanin content was calculated using the formula:

$$\text{Total anthocyanin (mg/100 g)} = \text{O.D} \times \frac{\text{Dilution factor}}{98.2} \times \frac{100}{w}$$

Where,

w = weight in gram

O.D. = absorbancy of samples

98.2 = extinction coefficient for cranberry anthocyanin

6.1.1.8 Estimation of ascorbic acid

6.1.1.8.1 Reagents

   a. 2, 6–dichloro Indophenol reagent

   b. 0.4% oxalic acid

6.1.1.8.2 Method

Ascorbic acid content was determined by the method of Mahadevan and Sridhar (1998). One gram of peeled root sample was homogenized with 10 ml of 0.4% oxalic acid and centrifuged at 5°C for 15 mins in a refrigerated centrifuge. A five ml extract was taken in an 100 ml beaker and titrated against the standardized sodium 2, 6 – dichloro Indophenol reagent until the solution become pink which persisted for at least 30 sec. The ascorbic acid content of the extract was calculated by using the formula:

$$i \times s \times d/a \times 100/w = \text{mg ascorbic acid/100 g roots}$$

Where,

   i = ml of Indophenol reagent used in the titration

   s = mg of standard ascorbic acid reacting with 1 ml of Indophenol reagent

   d = volume of extract in ml

   a = the aliquot titrated in ml

6.1.1.9 Estimation of phenol

6.1.1.9.1 Reagents

    a. 1% Folin–Ciocalteu reagent
    b. 20% Na_2CO_3 solution

6.1.1.9.2 Method

Phenolic in sample was estimated by Folin–Ciocalteu method (Mahadevan and Sridhar 1998). A ground freeze–dried root sample of 0.5 g was weighed and phenol was extracted with 50 ml of 80% aqueous on an ultrasonic bath for 20 mins. An aliquot (2 ml) of the extracts was centrifuged for 5 mins at 14000 rpm. One ml of the diluted sample was taken in a test tube to which 1 ml of Folin–Ciocalteu reagent followed by 2 ml of 20% Na_2CO_3 solution was added. The test tubes were shaken vigorously and boiled on water bath for 1 min. Thereafter, the test tubes were cooled and diluted to 25 ml in volumetric flasks. The absorbency of the sample was estimated at 650 nm and the phenolic content was calculated from a standard curve drawn with 0–100 μg/ml caffeic acid per assay.

6.1.2 Biochemical analysis of barley and malt

The parameters such as starch, crude protein and mineral contents of barley and malt were analyzed in a process identical to that of the analysis of purple sweet potato roots. The analyses of starch, crude protein and minerals are described in Section 6.1.1.3, 6.1.1.2, and 6.1.1.6, respectively. The methods of analyses of saccharose, reducing sugars and cellulose in the barley and malt are described below.

6.1.2.1 Estimation of saccharose

6.1.2.1.1 Reagents

    a. Sulphuric acid (H_2SO_4)
    b. Anthrone reagent
    c. Saccharose

6.1.2.1.2 Method

Malt and Barley were ground to fine materials. Samples were dried at 60°C for 4 hr. Samples weighing 20 mg were taken in a 15 ml Falcon tube. Four ml of water was added and thoroughly vortexed. The sample was kept in a water bath for 40 mins at 70°C. Again it was vortexed for 15 mins. After incubation it was kept in ice. Centrifugation was done at 3000 rpm for 10 mins and the supernatant was collected. 2.5 ml of the diluted sample was carefully transferred it to a new falcon tube in cold water. Five ml of anthrone reagent was carefully added to the tube in shaking condition kept in water and ice. After the addition of anthrone reagent the tubes were closed and vortexed in a water bath at ebullition for 7.5 minutes. The tubes were brought and cooled at room temperature. Absorbance was taken in 630 nm in a UV–Vis Spectrophotometer. Saccharose estimation was done by plotting the value of the absorbance of the sample against the standard curve prepared by 0, 6, 12, 18, 24, 30, 40, and 60 ug saccharose/ml dilutions.

6.1.2.2 Estimation of reducing sugar

6.1.2.2.1 Sample preparation

Barley and malt samples weighing 0.2 g were extracted at 50°C with 10 ml of distilled water in a water bath for 30 mins. The extracts were separated from the solid residue by centrifuging at 4000 rpm for 15 mins and transferred to a 25 ml volumetric flask. The same procedure was repeated twice with 10 ml and 5 ml of water for 15 mins each. The extracts were combined in a final 25 ml volume.

6.1.2.2.2 Reagents

a. Reagent A – 25 g Na_2CO_3 (anhydrous) + 25 g Na- K- tartrate + 20 g $NaHCO_3$ + 200 g Na_2SO_4 (anhydrous) were added to 1 litre distilled water. The reagent was stored overnight at room temperature before use.

b. Reagent B – 15 g of $CuSO_4$. $7H_2O$ was dissolved in 100 ml distilled water. Then 1 to 2 drops concentrated H_2SO_4 was added to prepare the reagent.

c. Reagent C – 1 ml reagent of B + 25 ml reagent A. This reagent was used fresh.

d. Arsenomolybdate reagent – 25 g of ammonium molybdate was mixed with 450 ml distilled water. To it, 21 ml concentrated H_2SO_4 and 0.3 g of sodium arsenate which was dissolved in 25 ml distilled water was added. This reagent was stored in brown bottle and was used within 24 hrs.

6.1.2.2.3 Method

The reducing sugar content of barley and malt was estimated by Nelson's method (Mahadevan and Sridhar 1998). Samples were taken in test tubes to which reagent C was incorporated. Test tubes were subjected to boiling in hot water bath for 20 mins and further cooled under running tap water. After cooling, 1 ml of arsenomolybdate reagent was added to the test tubes and a blue colour developed which was diluted to 25 ml in volumetric flasks. A blank was prepared in a similar manner but without barley and malt samples. The absorbency of the samples was recorded at 520 nm and readings were estimated from a standard prepared from 0–100 µg/ml D–glucose per assay.

6.1.2.3 Estimation of cellulose

6.1.2.3.1 Reagents

a. Acetic/Nitric reagent: 150 ml of 80% CH_3COOH and 15 ml of concentrated HNO_3 was mixed.

b. Anthrone reagent: 200 mg anthrone was dissolved in 100 ml concentrated H_2SO_4.
It was prepared fresh and chilled for 2 hrs before use.

c. H_2SO_4 (67%).

6.1.2.3.2 Method

Acetic/nitric reagent of 3 ml was added to a 0.5 g of sample in a test tube. It was properly mixed in a vortex mixer. The test tube was subjected to boiling at 100°C using a water bath for half an hour. The content was centrifuged for 15 to 20 mins. Supernatant was removed and the remains were washed with distilled water. 10 ml of 67% sulfuric acid was added to it and it was kept for 1 hr. 1 ml of the above solution was diluted to 100 ml. To 1 ml of the diluted solution 10 ml of anthrone reagent was mixed properly. Boiling water was used to heat the tubes for 10 mins. The solution was cooled and the OD (Optical Density) was recorded at 630 nm. The concentration of cellulose was measured by plotting the absorbance of the sample solution in the standard curve (40–200 mg cellulose/ml).

6.1.3 Biochemical analysis of must, wine, wort and beer

The procedure of biochemical analysis for various parameters used for must and wine as well as for wort and beer are described in details in the following sections.

6.1.3.1 Estimation of total sugar, starch and reducing sugar

The analyses procedure of total sugar and starch were identical to that described for purple sweet potato roots in the Section 6.1.1.3. The diluted sample was directly added to the anthrone reagent instead of the alcoholic extract that was added in case of purple sweet potato roots. Reducing sugar was estimated identically to that of barley and malt described in Section 6.1.2.2.

6.1.3.2 Estimation of total phenols

The analytical procedure for estimation of total phenolic content was identical to the procedure applied for purple sweet potato roots (Section 6.1.1.9). Diluted sample of 1 ml was directly analyzed for the estimation of phenols instead of extraction procedure (that was applied for purple sweet potato roots).

6.1.3.3 Estimation of anthocyanin

The procedure of anthocyanin estimation in the must, wine, wort and beer was similar as described for the purple sweet potato roots. Samples (must, wine, wort and beer) were directly used for the analysis method as described in Section 6.1.1.7.

6.1.3.4 Determination of pH

Must, wine, wort and beer samples were properly shaken before taking the pH readings. The reading was carried out by using the glass electrode of a pH meter (Mahadevan and Sridhar 1998). Prior to the pH recording, the pH meter was standardized using buffer solutions of pH 7 and 9.2.

6.1.3.5 Estimation of titratable acidity

Amerine and Ough (1980) were followed for the evaluation of titratable acidity. All the samples (must, wine, wort and beer) in their liquid state were used as it were for the evaluation of titratable acidity.

6.1.3.5.1 Reagent

0.1 N NaOH

6.1.3.5.2 Method

Five ml of the sample was put into a 100 ml beaker and titration was conducted against 0.1 N NaOH. Phenolphthalein was used as an indicator and the end point of the titration was observed as the sample turned pink in colour that persisted for half a minute. The titratable acidity of the sample was computed as per the formula:

$$\text{Titratable acidity } (g/ml) = \frac{v_1 \times n \times 75 \times 100}{1000 \times v_2}$$

Where, v_1 = volume of NaOH solution used for titration

n = normality of NaOH solution

v_2 = volume of the sample

6.1.3.6 Estimation of lactic acid

6.1.3.6.1 Reagents

a. 20% $CuSO_4$ solution
b. 4% $CuSO_4$ solution
c. 1.5% Para–hydroxydiphenyl solution

6.1.3.6.2 Method

Sample volume of one ml was put in a test tube. To the sample bearing test tube seven ml of distilled water along with 20% $CuSO_4$ pentahydrate solution of one ml was added. Further, one g of $Ca(OH)_2$ was mixed with the solution in the test tube. A fine suspension was formed by shaking the mixture thoroughly. Then the suspension was kept without disturbing for 30 min. The solution was further centrifuged at 4000 rpm for 15 min and further one ml of the supernatant was brought to a borosilicate test tube. To the supernatant bearing test tube 0.05 ml of 4% $CuSO_4$ solution was mixed to it followed by six ml of concentrated sulfuric acid. The solution bearing test tube was permitted for boiling in a hot water bath for 7 mins and again put in a ice bath for bringing down the temperature bellow 20°C. After the test tubes are cooled, para–hydroxydiphenyl solution (1.5%) of 0.1 ml was mixed into the solution resulting in precipitation. The precipitate bearing test tubes were shaken vigorously to dissolve the precipitate. Further, the test tube was kept at 30°C for 15 mins over a water bath till the development of purple colour. In a similar manner a control solution was prepared without taking the sample. The absorbency of was recorded at 560 nm in an UV–VIS Spectrophotometer (Cecil Instruments, Cambridge, UK). Lactic acid content was estimated using a standard curve plotted with 0– 70 µg/ml of lithium lactate per assay.

6.1.3.7 Estimation of total soluble solids (TSS)

The TSS of the samples in liquid state (must, wine, wort and beer) was measured by using a hand refractometer (Sipcon, Jalandhar, Punjab, India).

6.1.3.8 Estimation of tannin

6.1.3.8.1 Reagents

a. Buffer solution, pH 7.9
    In a 500 ml volumetric flask 1.36 g of potassium dihydrogen phosphate, 8.35 g of disodium hydrogen phosphate and 12.5 g of sodium hydrogen carbonate were added to water and was diluted to volume (500 ml).

b. Cinchonine sulphate solution
    Into a 100 ml volumetric flask 1.5 g of cinchonine base was dissolved in 2 ml of 1:3 sulphuric acid–water and further diluted to a volume of 100 ml with distilled water.

c. Ethanolic hydrochloric acid solution

Into a 100 ml volumetric flask 10 ml of concentrated HCl was added and diluted to 100 ml by 95% ethanol.

6.1.3.8.2 Method

Fifty ml of the samples was placed into a 100 ml centrifuge tube, and neutralized to pH 7.0 with 1 N Sodium hydroxide solution. To the mixture 25 ml of buffer solution (pH 7.9) and 12.5 ml of the cinchonine sulphate solution were added. The solutions were mixed and allowed to stand for 20 mins at room temperature. The solution was centrifuged to remove the precipitate. The clear supernatant was removed into a 200 ml volumetric flask. The precipitate was washed twice with 10 ml of 10% sodium sulphate solution and the wash was combined with the supernatant. The combined wash and supernatant were acidified to pH 3.5 with 1 N hydrochloric acid and the volume was made up to 200 ml. The precipitate was dissolved in 50 ml of ethanolic hydrochloric acid solution. The total phenol content of both solutions were measured and reported as caffeic acid. The solution made from the precipitate provided the amount of hydrolysable tannin phenols while the results of combined supernatant and wash provided the other phenolic compounds.

6.1.3.9 Estimation of DPPH scavenging activity

6.1.3.9.1 Reagents

   a. 2,2–diphenyl–1picryl hydrazyl
   b. Methanol
   c. Butylated hydroxytoluene

6.1.3.9.2 Method

The free radical scavenging effect of the solvent extracts of selected samples were determined by DPPH radical scavenging assay following the method of Hatano et al. (1988). Briefly, 2.0 ml of 0.1 mM DPPH (2, 2–diphenyl–1–picrylhydrazyl) solution (in methanol) was added to the test tubes containing 0.1 ml aliquot of solvent extracts of selected must, wort, wine and beer samples and standard BHT (Butylated hydroxytoluene) at 50, 75 and 100 µg/ml concentration. The mixture was vortexed for 1 min and kept at room temperature for 30 min in the dark. The absorbances of all the sample solutions were measured at 517 nm. The percentage scavenging effect was calculated from the following equation.

   % DPPH scavenging activity = $[(A_0 - A_1)/A_0] \times 100$

Where A_0 = Absorbance of control
   A_1 = Absorbance of test sample

6.1.3.10 Determination of specific gravity

The specific gravity of the samples of must, wine, wort and beer was measured by a hydrometer with a range of 1.000 to 1.099 (Amerine and Ough 1980).

6.1.3.11 Estimation of ethanol content

The ethanol content of the wine and beer samples was measured by the difference of specific gravity of the wine and beer before and after fermentation. The difference between the specific gravity reflects the alcohol content as per the recommendation of Palmer (2006).

6.2 Technoeconomical Feasibility and Cost Economics Study for Industrial Scale Production

An assumption is proposed for an integrated winery cum brewery to be established in the innovative technologies developed for the preparation of purple sweet potato wine and purple sweet potato beer. An integrated production unit of 2500 litres purple sweet potato wine/day and 7500 litres purple sweet potato beer/day was assumed. The schematic design of the proposed winery and brewery is presented in Fig. 6.1 and Fig. 6.2. The study includes the total capital, fixed assets, working capital, means of finance, schedule of production, sale of purple sweet potato wine and beer, calculation of manufacturing cost per unit of purple sweet potato wine and beer, projected profitability statement and details of loan. In this study effort has been given to

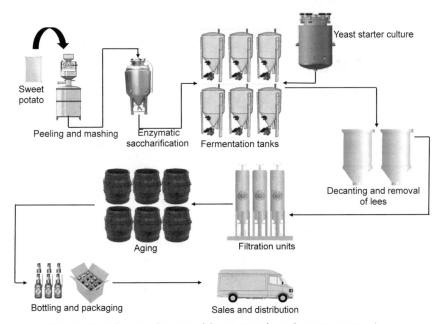

Figure 6.1. Schematic diagram of the winery of purple sweet potato wine.

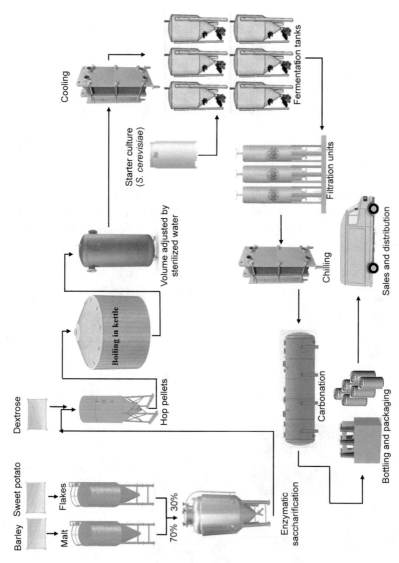

Figure 6.2 Schematic diagram of the brewery for production of purple sweet potato beer.

keep costing values as realistic as possible. The values of different prices and bank interests have been enquired of before using in the projection. Hence, most of the values are realistic and not imaginary.

7

Results and Discussion

Biochemical, Sensory, Statistical and Infrared Studies

Alcoholic fermentation of fruits, vegetables and cereals are carried out as one of the most important technologies used to preserve and store finished and partially-finished food products. Studies are being conducted with the aim of improving the quality of alcoholic beverages keeping in view the demands of the products as well as the quantity of raw materials used in the beverage industry. The prime reasons attributed to the interest of the research in this industry include nutritional, physiological and the hygienic aspects of the procedure. Due to the agricultural, nutritional, sensory and preservation reasons, alcoholic fermentation is used as a prospective technology for the bio-processing of sweet potato. Sweet potato roots are high in starch content, sugars, vitamin C, pro-vitamin A, iron and some varieties are known to contain coloured pigments such as β-carotene and anthocyanin (Yamakawa 1997). These pigments have the following benefits:

- They act as anti-oxidants, anti-cancer and anti-ageing compounds
- They provide protection against night blindness and liver injury
- They alleviate vitamin A deficiency, cataract and macular degeneration

In the present study, a variety of sweet potato rich in anthocyanin (Fig. 1.2) has been processed for alcoholic fermentation. This is because of its higher content of anthocyanin pigments and phenolics which work as antioxidants. In comparison to β-carotene the anthocyanin pigments are more stable at higher temperature (>70°C). In the preparation of wine and beer from sweet potato, saccharification at higher temperature (>70°C) is necessary for converting the sweet potato starch into sugars. Hence, the variety containing anthocyanin is preferred over the β-carotene rich variety.

Three novel fermented food products were developed from the sweet potato variety rich in anthocyanin by alcoholic fermentation such as:

- Anthocyanin rich wine (red wine)
- Herbal wine
- Anthocyanin–rich beer

7.1 Biochemical Composition of Purple Sweet Potato Roots

The biochemical composition of the purple sweet potato has been presented in the following Table (7.1). From the table, it is evident that this variety is rich in phenolics and anthocyanin pigments (both considered to be antioxidants) as well as in starch, besides containing an array of macro and micro-nutrients. This variety is therefore chosen for making wine and beer.

7.2 Biochemical Composition of Barley and Malt

The biochemical composition of the barley and malt has been displayed in Table 7.2. Malt is used as substrate for the preparation of beer as demonstrated in the section Materials and Methods (Section 5.7). Malt was prepared from barley by soaking, draining, germinating and heating (in an oven) the barley grains. The malting process has been described in Section 5.7.2.1 of the Materials and Methods.

Table 7.1. Biochemical composition of purple sweet potato used in this study.

Constituents		Sweet potato mean value
Moisture (g/kg)	:	630.0 ± 2.56
Protein (g/kg)	:	14.0 ± 1.15
Starch (g/kg)	:	178.0 ± 2.69
Total sugar (g/kg)	:	22.8 ± 0.31
Dietary fibre (g/kg)	:	16.0 ± 0.25
Ash (g/kg)	:	7.1 ± 0.14
Ca(mg/100 g dwb)	:	290 ± 0.19
P(mg/100 g dwb)	:	510 ± 0.09
Mg(mg/100 g dwb)	:	260 ± 0.38
Na(mg/100 g dwb)	:	520 ± 0.26
K(mg/100 g dwb)	:	2600 ± 1.23
S(mg/100 g dwb)	:	130 ± 1.15
Fe(mg/100 g dwb)	:	4.9 ±0 .3
Cu(mg/100 g dwb)	:	1.7 ± 0.3
Zn(mg/100 g dwb)	:	5.9 ± 0.2
Mn(mg/100 g dwb)	:	1.1 ± 0.3
Al(mg/100 g dwb)	:	8.2 ± 0.2
B(mg/100 g dwb)	:	1.0 ± 0.2
Anthocyanin (mg/kg)	:	740–800 ± 0.52
Ascorbic acid (g/kg)	:	240 ± 0.45
Phenol(mg/100 g)	:	65

Table 7.2. Biochemical composition of barley and malt.

Composition of barley and malt (% d.m.)		
Composition	Barley	Malt
Starch	63–65	58–60
Saccharose	1–2	3–5
Reducing sugars	0.1–0.2	3–4
Cellulose	4–5	5
Raw proteins	8–11	8–11
Minerals	2	2.2
Other compounds	9–17	13–21

7.3 Anthocyanin Rich Purple Sweet Potato Wine (Red Wine)

Wine is traditionally prepared from grapes (*Vitis vinifera*), however, several ripe tropical and subtropical fruits have been employed for the successful production of wine. For example mango, litchi, kiwi, banana, and cashew apple have been found appropriate for the preparation of wine as they contain levels of high sugar and water, an appropriate taste, flavour and are widely available (Muniz et al. 2008). In the current research the traditional wine making procedure has been deviated, in place of grapes and ripe tropical fruits, a starch rich substrate, purple sweet potato has been used instead.

7.3.1 Saccharification of purple sweet potato roots

7.3.1.1 Application of enzymes

Purple sweet potato roots were saccharified by the application of the amalgamation of two commercial enzymes, Termamyl [thermostable (>90°C) α-amylase enzyme] and Dextrozyme [amyloglucosidase (glucoamylase) (>90°C)] described earlier in the Materials and Methods Section 5.4.

Termamyl was applied to homogenized purple sweet potato roots at a dose of 0.2% and incubated at 90°C for 1 hr. The liquefied mash was cooled down to 45°C and 1% Dextrozyme G.A was applied to it and it was further incubated for 48 hrs at 45°C for saccharification. The details of saccharification process of purple sweet potato roots for wine making are described in Materials and Methods (Section 5.5.1.1). In this step, the saccharification, i.e., the conversion of starch to glucose was found to be 98.2% which indicates that the original starch content, 174.85 g/kg in the roots was converted to 176.2 g net total sugar/kg root.

Numerous studies have reported on the application of both Termamyl and AMG in conjugation with the saccharification of starch to reducing sugars. Govindaswamy et al. (1997) showed the saccharification of cassava sago pearl varied from 32–98% in different physico-chemical conditions when applied with Termamyl and AMG. Similarly, conversion of 97% of starch in algal biomass was observed by Choi et al. (2010a), in which *Chlamydomonas reinhardtii* Utex 90 was saccharified by the application of the same two enzymes for further applications in the manufacture of bioethanol. Furthermore, it has been reported that the percentage of conversion of cassava starch to dextrin, i.e., the liquefaction process in high fructose syrup production was better at 90°C for 1hr Termamyl treatment than at 80°C for 2 hrs (Paolucci et al. 2000). Similarly, the combined application of Termamyl at 90°C for one hr and AMG at 45°C for 24 hrs could convert 85% of starch in sweet potato flour and roots into sugar for successive conversion into fuel ethanol (Ray and Naskar 2008).

With this context, our results on saccharification of purple sweet potato roots by Termamyl+AMG corroborate these findings.

7.3.2 Stock culture

The microorganism used in the preparation of purple sweet potato wine was *S. cerevisiae*. The microorganism was preserved in our laboratory in PDA slants at 4°C. The detailed information about the organism has been described in the Materials and Methods (Section 5.3).

7.3.3 Starter culture

Starter culture of *S. cerevisiae* for the preparation of purple sweet potato wine was prepared by using grape juice and has been described in the Materials and Methods (Section 5.5.1.2.).

7.3.4 Wine making process

The steps in wine making involve the primary processing of purple sweet potato, i.e., washing, de–skinning and crushing with tap water.

After primary processing the other steps were saccharification, juice extraction, amelioration of must, fermentation, racking and bottling. The entire process of wine making has been described in Section 5.5.1 of the Materials and Methods.

7.3.5 Biochemical analyses

The changes in parameters, i.e., TSS, starch, total sugar, titratable acidity, pH, phenolic content, anthocyanin, tannin, lactic acid, ethanol and DPPH scavenging from must to wine have been studied by the procedure described in the Materials and Methods (Section 6.1.3).

The composition of must and wine is shown in Table 7.3. Must in the study was the ameliorated juice from the saccharified purple sweet potato root before fermentation and wine was the undistilled alcoholic beverage made from the must after fermentation.

Table 7.3. Biochemical composition of purple sweet potato must and wine and changes in the mean proximate variables between the must and wine (paired '*t*' test).

Parameter	Must	Wine	Mean difference	'*t*' value
TSS(°Brix)	20.35	2.25	18.10	271.52**
Starch(g/100 ml)	3.15	0.15	3.00	116.77**
Total sugar(g/100 ml)	14.49	1.35	13.14	184.33**
Titratable acidity (g tartaric acid/100 ml)	0.54	1.34	−0.80	−149.77**
pH	5.24	3.61	1.63	189.29**
Phenol[caffeic acid equivalent (CAE) g/100 ml]	0.37	0.36	0.01	2.61**
Anthocyanin(mg/100 ml)	65.05	55.09	9.96	12.06**
Tannin(mg/100 ml)	0.74	0.64	0.10	13.25**
Lactic acid (mg/100 ml)	0.01	1.14	−1.13	−302.58**
Ethanol (% v/v)	0	9.33	−9.33	−426.32**
DPPH(%)	72.75	58.95	13.80	40.92**

**Significant at 0.01% level (two-tailed).

7.3.5.1 Acidity

The titratable acidity (TA) rose from 0.54 (g tartaric acid/100 ml) to 1.34 (g tartaric acid/100 ml) in the wine. The raise of TA was associated with a decrease in pH from 5.24 to 3.61 in wine. The must had a lactic acid content of 0.01 mg/100 ml which increased to a value of 1.14 mg/100 ml.

7.3.5.2 Carbohydrates

As per anticipation the total sugar content (g/100 ml) fell from an initial value of 14.49 in must to 1.35 in wine. The TSS of the must was 20.35°Brix which decreased to a value of 2.25°Brix in wine. The starch content of the must and wine were 3.15 g/100 ml and 0.15 g/100 ml, respectively. Comparable findings were obtained in the fermentation of jamun (*Syzygium cumini* L.) (Chowdhury and Ray 2007), cashew apple (*Anacardium occidentale* L.) (Mohanty et al. 2006) and mango (*Mangifera indica* L.) (Reddy et al. 2005) fruits into wine.

7.3.5.3 Antioxidants

The anthocyanin concentration in must decreased from 65.05 mg/100 ml to 55.09 mg/100 ml in wine. Other fruit wine also showed similar pattern of result. In jamun wine, the anthocyanin concentration was 60 mg/100 ml, which was comparable to that of grape wine (60–67 mg/100 ml) (Chowdhury and Ray 2007). Boiled and non-boiled purple sweet potato roots were employed for the production of lacto–juice which showed the anthocyanin concentration of 97.1 and 160 mg/100 ml, respectively (Panda et al. 2009b). The wine was recorded with a DPPH scavenging activity of 58.95% whereas in case of must it was higher, i.e., 72.75%. Choi et al. (2010b) revealed the DPPH radical scavenging activity for 5% purple sweet potato extract blended fermented soy milk. The 5% purple sweet potato extract–added fermented soy milk could scavenge 2, 2–diphenyl – 1 picryl hydrazyl to 70.5% at a dose of 500 µg/ml. Suda et al. (2008) reported an intake of purple sweet potato beverage, rich in acylated anthocyanins (400.6 mg anthocyanins/

250 ml beverage) could considerably reduce the serum levels of hepatic biomarkers, particularly the serum γ-glutamyl transferase level, in healthy men (30–60 years) with borderline hepatitis. However, there was no significant variation in phenolic content in the purple sweet potato must and wine in our study (Table 7.3). The total phenolic content of the must and wine were 0.37 g/100 ml and 0.36 g/100 ml, respectively. The phenolic contents of sweet potato vary widely ranging from 0.14–9.45 mg of chlorogenic acid equivalent/g fw (Teow et al. 2007, Cevallos–Casals and Cisneros–Zevallos 2003). Genetic features and growing conditions play an important role in the formation of secondary metabolites including phenolic acids (Ray and Tomlins 2010). The tannin content of the must was 0.74 mg/100 ml and that of the wine was 0.64 mg/100 ml.

7.3.5.4 *Alcohol content*

Few studies have been carried out on fermented food products produced from sweet potato. Wireko–Manu et al. (2010) produced a non-alcoholic beverage from sweet potato with pH 3.81 to 4.34, TSS, 12.0 to 13.1°Brix and TA ranged from 0.45 to 1.6 (g/l tartaric acid), with lime flavoured. Teramoto et al. (1998) re-produced a popular traditional Japanese alcoholic beverage, *miki* using cooked rice and mashed raw sweet potato roots as the saccharifying agent in different combinations (a mixture of 30 g of ground autoclaved rice with 6, 12, 30 and 60 g of raw sweet potato mash, respectively). The reproduced miki had an alcohol content of 5–6% (v/v); it had a pleasurable aroma with a sour taste and weak bitterness. Oligosaccharides including maltotriose and maltopentaose were identified in *miki* prepared with greater proportion of sweet potato mash. Similarly, a rice beverage incorporating sweet potato was prepared by adding barley sprouts, sweet potato or a mixture of barely sprouts and sweet potato (1:1). Amylases from barely sprouts and sweet potato had a similar hydrolysis pattern to β-amylase. The use of sweet potato could enhance the sweetness, flavour, and improved fondness in rice beverage (Suh et al. 2003). It was also observed that the beverages developed from raw purple sweet potato roots have a larger amount of deacylated anthocyanins as compared to the beverages produced from the cooked purple sweet potato (Saigusa et al. 2005). Filtrate volume and ethanol concentration of the beverage manufactured from raw purple sweet potato was found to be more than that of the cooked purple sweet potato (Saigusa et al. 2005).

7.3.6 FTIR analysis

FTIR spectroscopy is a promising technique to provide information on the antioxidant potential of red wine (Versari et al. 2010). In FTIR, the wave numbers selected by the mathematical algorithm are associated with the absorption bands of the most important compounds in wine. In the mid FTIR region, the main ranges selected (3,700–3,000, 1,700–1,400 and 1,200–1,000/cm) correspond to the absorption bands of O–H, C=O and C–C/C–O stretching vibration (Colthup et al. 1990).

Samples of purple sweet potato wine and a commercial grape wine were subjected to FTIR analysis for detection and comparison of functional groups (Fig. 7.1) as described in the Materials and Methods (Section 5.5.3). The spectra (0–5000/cm) were interpreted by using the guidelines from Stuart (2004).

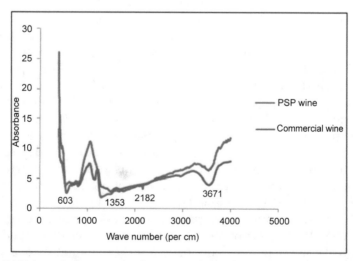

Figure 7.1. Infra–red spectrum of anthocyanin rich purple sweet potato (PSP) wine and a commercial wine.

The strong phenol O–H stretch detected at 3671/cm was indicative of the presence of alcohol and phenol groups in both purple sweet potato and commercial wine. Similarly peaks detected at 1843/cm, 1495/cm, 1179/cm and 773/cm showed the presence of anhydrides group (C=O stretching), Azo compound (N=N stretching), second overtone C–H stretching and alkynes (=C–H bending) respectively in the commercial wine sample. Further the C–N stretching bands showed the presence of amines. Aromatic amines showed bands in the range of 1323–1250/cm and aliphatic amines showed bands at 1220–1000/cm in both the samples. Primary (–NH2), secondary (–NH) and tertiary (no hydrogen attached to N) amines may be differentiated by using the infrared spectra. Some additional new bands were detected in the case of purple sweet potato wine, alkynes were detected in form of C=C stretching at 2182/cm and =C–H bending at 603/cm. A prominent peak was observed at 1353/cm which represents aromatic C–N stretching compounds.

A comparable study was conducted by Versari et al. (2010) to predict the total antioxidant capacity of red wine using FTIR spectra in a range of 600 to 4400/cm. The result revealed a good correlation between phenols and the antioxidative property of red wines (Versari et al. 2010). Another study was conducted to determine the quantitative value of methanol by using IR spectrometry. Spectral measurements were done for 10 samples in a wavelength region of 4000–600 cm^{-1}. The study predicted the methanol concentration (0.5–2.5%) using the developed calibration model (Nagarajan et al. 2006).

7.3.7 Statistical analysis

Multivariate statistical analysis was performed by using statistical software SPSS (SPSS software for windows release 17.0 SPSS, Chicago, IL, USA). Paired '*t*' test was used to identify the mean differences in the analytical variables of the must and the wine. PCA was applied to the analytical variables of the purple sweet potato wine to the group and the related variables into smaller components for interpretation of their relationship. Correlation was analyzed to identify the correlation between the variables.

Independent 't' test was applied to the organoleptic parameters to identify the mean difference of the sensory attributes. The details of these procedures are described in the Materials and Methods (Section 5.5.5).

7.3.7.1 Correlation of the analytical and proximate data of purple sweet potato wine

The correlation analyses among the proximate and analytical variables are presented in Table 7.4. The most important correlations were starch–TSS (0.739), starch–total sugar (0.619), TA-pH (–0.595), lactic acid–TA (0.925) and lactic acid – pH (–0.610).

Similar results were obtained during the preparation of lacto–pickles from β-carotene rich sweet potato. Significant correlations were observed between TA–pH (0.855), lactic acid–TA (0.974), lactic acid–pH (–0.801) and starch–total sugar (0.689) among the analytical variables analyzed in the pickle (Panda et al. 2007).

Correlation was also applied for the analytical variables of litchi wine produced from litchi fruits. Higher significant correlations observed were TSS–TA (–0.866), TA–pH (0.925), TA–phenol (0.866), phenol–tannin (0.786), etc. (Kumar et al. 2008).

7.3.7.2 Principal component analysis (PCA)

PCA was applied to reduce the 11 original analytical variables into four principal components (PC1–PC4), having eigen values larger than 1 and retained for rotation (Hair et al. 1998). PC1, PC2, PC3 and PC4 accounted for 30.27, 16.53, 16 and 13.85%, respectively and the total variations (76.65%) are shown in Table 7.5.

When the dimension was interpreted, the factor pattern was rotated using the varimax method. Basing on the report of Stevens (1992), an attribute was correlated to load heavily on a given component if the factor loading was greater than 0.72. All the 11 analytical variables showed strong loading on the dimension. Three analytical variables, i.e., TA (+ve), pH (–ve), and lactic acid (+ve) were strongly loaded

Table 7.4. Correlation coefficient for analytical+ proximate variables of purple sweet potato wine.

TSS	Starch	Total sugar	Titratable acidity	pH	Phenol	Anthocyanin	Tannin	Lactic acid	Ethanol	DPPH
1	0.739**	0.449*	0.257	-.0114	0.490*	0.015	0.211	0.131	0.102	0.090
	1	0.619**	0.483*	-0.471*	0.319	0.042	-0.015	0.441	0.346	-0.046
		1	0.067	-0.156	0.187	-0.262	-0.324	0.045	-0.043	-0.294
			1	-0.595**	0.239	-0.057	-0.108	0.925**	-0.005	0.292
				1	-0.089	-0.201	0.107	-0.610**	-0.036	0.175
					1	-0.310	0.096	0.147	0.092	0.449*
						1	0.212	-0.151	0.159	0.034
							1	-0.168	0.188	0.197
								1	-0.026	0.306
									1	0.013
										1

**Significant at 0.01 level (two-tailed).

105

Table 7.5. Principal components analysis of proximate variables (purple sweet potato wine).

Parameter	PC1	PC2	PC3	PC4
TSS	0.076	0.821	0.264	0.200
Starch	0.445	0.836	−0.037	0.205
Total sugar	0.017	0.801	−0.293	−0.342
Titratable acidity	0.912	0.129	0.246	−0.064
pH	−0.802	−0.166	0.264	−0.139
Phenol	0.048	0.392	0.729	−0.125
Anthocyanin	0.105	−0.195	−0.251	0.757
Tannin	−0.210	−0.021	0.387	0.729
Lactic acid	0.932	0.046	0.210	−0.150
Ethanol	−0.011	0.154	0.017	0.723
DPPH	0.164	−0.171	0.856	0.068
Variance explained (%)	30.27	16.534	16	13.845
Eigen value	3.33	1.819	1.76	1.523

on PC1 indicating strong correlation between the variables and can be regarded as "acid" axis. Total soluble solid (+ve), starch (+ve) and total sugar (+ve) have been loaded substantially on PC2 which can be inferred as "carbohydrate" axis. Phenol (+ve) and DPPH (+ve) were loaded on PC3. Anthocyanin (+ve), tannin (+ve) and ethanol (+ve) were loaded on PC4. Plotting PC1 loading against PC2 loading, PC1 vs PC3, PC1 vs PC4, PC3 vs PC4 and PC2 vs PC4 depicted these relationships visually (Fig. 7.2a, b, c, d and e, respectively).

Several studies have been conducted to reduce the analytical variables of fermented food products into smaller components by using PCA. In the preparation of β-carotene rich sweet potato curd PCA was applied to the 14 analytical variables. PCA reduced the variables into four principal components, i.e., PC1–PC4. PC1 accounted for 60% where as PC2, PC3 and PC4 accounted for 17, 12 and 7%, respectively. Seven analytical variables, i.e., β-carotene (−ve), ascorbic acid (−ve), calories (+ve), moisture (+ve), protein (+ve), fat (+ve) and dietary fibre (−ve) were loaded heavily on PC1. Substantial factor loadings of TA (+ve) and total sugar (+ve) were loaded on PC2. Organic matter

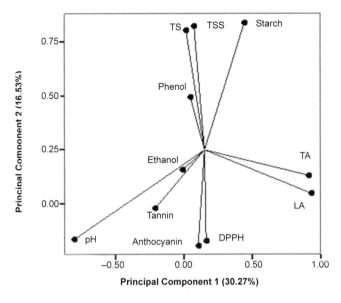

Figure 7.2a. Graphical representation of principal components PC1 vs. PC2 of analytical variables of purple sweet potato wine.

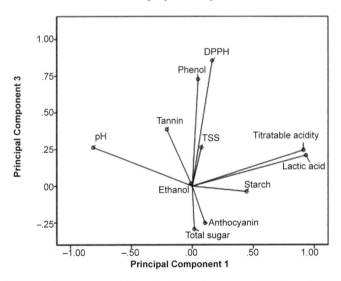

Figure 7.2b. Graphical representation of principal components PC1 vs. PC3 of analytical variables of purple sweet potato wine.

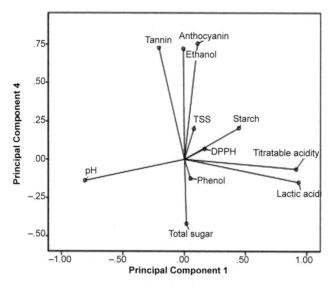

Figure 7.2c. Graphical representation of principal components PC1 vs. PC4 of analytical variables of purple sweet potato wine.

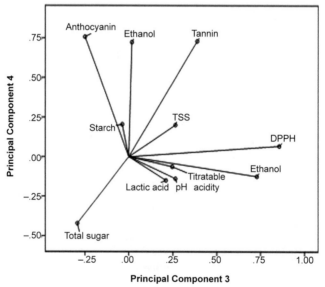

Figure 7.2d. Graphical representation of principal components PC3 vs. PC4 of analytical variables of purple sweet potato wine.

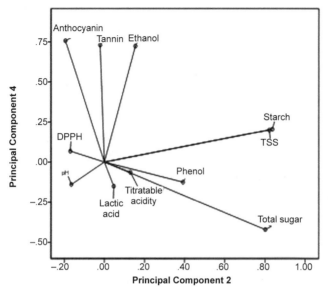

Figure 7.2e. Graphical representation of principal components PC2 vs. PC4 of analytical variables of purple sweet potato wine.

(+ve), starch (+ve), LA (−ve) and ash (−ve) were loaded on PC3. pH was loaded on PC4 (Mohapatra et al. 2007).

Another study was carried out on lactic acid fermented cucumber juice and the analytical variables were reduced to two independent principal components, i.e., PC1 and PC2. PC1 accounted for 71.6% whereas PC2 accounted for 16.5% of the total variations (Kohajdova et al. 2007).

7.3.8 Sensory evaluation

Data presented in Table 7.6 showed the reference grape wine being superior when clarity ($t = -6.11$, $p < 0.01$), colour ($t = -5.81$, $p < 0.01$), odour ($t = -9.02$, $p < 0.01$), flavour ($t = -7.46$, $p < 0.01$), spritz ($t = -15.17$, $p < 0.01$) and finish ($t = -13.65$, $p < 0.01$) were taken into account. However, there was no difference found in viscosity. Although the reference wine was finer than purple sweet potato wine, but all the mean differences of the organoleptic attributes was less than 1, which revealed that the wines differed only within an extent of 10%, hence it might be conceded that the sensory quality of purple sweet potato wine is comparable with the commercial grape wine.

Earlier studies by Chowdhury and Ray (2007) produced wine from jamun fruits and compared it with that of commercial grape wine. The sensory evaluation analysis showed that the jamun wine was little inferior to that of the commercial grape wine (except colour and appearance) (p < 0.05) but attributes like aroma, taste, aftertaste and colour/appearance scored about 3.0 (like much).

Similarly, sensory analysis has been applied to standardize the preparation of sweet potato patties. The evaluation indicated that patties of acceptable quality could be prepared from either freshly harvested sweet potato or from roots that had been cured and stored for upto 6 months (Hoover et al. 1983)

Anthocyanin rich sweet potato wine was a reddish pink coloured beverage (Fig. 7.3) with an alcohol content of 9.33%.

Table 7.6. Attribute-wise mean differences and '*t*' values for purple sweet potato wine and grape wine.

Attributes	Sensory scores of wine samples					
	Purple sweet potato wine		Grape wine			
	Mean	RSD*(%)	Mean	RSD(%)	Mean difference	'*t*' value
Clarity	6.99	5.68	7.36	8.11	−0.38	−6.11**
Colour	7.23	5.58	7.55	6.91	−0.32	−5.81**
Viscosity	6.90	6.85	6.93	4.18	−0.04	−0.75[NS]
Odour	7.47	4.76	7.85	5.00	−0.38	−9.02**
Flavour	7.55	5.46	7.93	6.39	−0.39	−7.46**
Spritz	6.88	7.24	7.68	5.05	−0.81	−15.17**
Finish	7.02	5.28	7.56	4.00	−0.54	−13.65**

*RSD – Relative standard deviation.
**Significant at 0.01 level (two-tailed).

Figure 7.3. Anthocyanin rich purple sweet potato wine.

7.4 Herbal/Medicinal Wine from Purple Sweet Potato

Curative herbs and spices have been used in food since the ancient times, not only as agents of flavour, but also as folk medicine and food preservatives (Beuchat 1994, Cutler 1995).

The herbs possess important properties such as anti-oxidant and bacteriostatic or bactericidal activity that enhances the shelf–life of the food and beverages (Beuchat and Golden 1989) and it also enriches the food by imparting specific phytonutrients, vitamin and mineral. Herbal or medicinal wines are the fermented alcoholic beverages produced from fruits or medicinal plant parts which have pharmaceutical and therapeutic properties. There are several studies being conducted to produce a wine having medicinal properties. Soni et al. (2009)

illustrated the bioprocessing of Indian gooseberry (*Emblica officinalis*) fruits having curative properties for example as a diuretic, laxative, carminative, stomachic, astringent, anti-diarrhoeal, anti-haemorrhagic and anti-anaemic to wine. *Aloe vera* is known to have many medicinal properties and has also been fermented to wine (Pongparnchedtha and Suwanvisolkij 2011). Wine has been prepared from tea (*Cammellia sinensis*) leaves as substrate, having well accepted sensory and health characteristics (Aroyeun et al. 2005). Bhat and Moskovitz (2009) demonstrated the preparation of herbal tea by mixing indigenous South African medicinal herbs known for combating arthritis, constipation, and hangover, cardiac and other diseases along with tea and promoted. However, in all these studies, the fermentation process was very slow as these plants are low in starch content; conversely the ethanol concentrations were below 5% (v/v). It is therefore, suggested either to add cane sugar into the fermentation medium to accelerate the process or alternatively we may use starchy crops like purple sweet potato as the source (substrate) and add medicinal plants to the must to develop medicinal wine.

In our previous study a red wine (anthocyanin rich) was developed from purple sweet potato (Section 7.3). In continuation of study, a herbal wine was developed from purple sweet potato roots by fortifying the saccharified pulp with selected medicinal plant parts (Table 5.1) and further fermented with *S. cerevisiae*.

7.4.1 Saccharification of purple sweet potato roots

The enzymes used in the sacchrification of purple sweet potato roots were the same as those used in the preparation of red wine from purple sweet potato roots (Section 7.3.1). The conversion of starch into sugars was 97%. From the original content of 178 g/kg starch in the roots, the total sugar generated after saccharification was 189.92 g/kg roots. A similar result of conversion (98.2%) was obtained in the case of red wine from purple sweet potato (Section 7.3.1).

7.4.2 Stock culture and starter culture

The stock culture used for the preparation of the starter culture was the same as the culture used for creating red wine from the purple sweet potato (Section 7.3.2). The process of the starter culture preparation was identical to that of the purple sweet potato wine (red wine) described in Section 7.3.3.

7.4.3 Wine making process

This process has been deviated from the general perception of herbal or medicinal wines. Medicinal plant parts contain a lesser amount of sugar so the yield of ethanol is less in the wines prepared by directly fermenting the plant parts, which is a drawback in the process of wine making. Instead of directly fermenting medicinal plant parts to wine, selected medicinal plant parts (Table 5.1) were taken as an adjunct along with the saccharified pulp of purple sweet potato (described in Section 7.3.1) to add medicinal attributes as well as to enhance the ethanol concentration in the wine. The details of the wine making process have been described in the Section (5.6.2) of the Materials and Methods.

7.4.4 Biochemical analyses

The studied parameters were TSS, reducing sugar, starch, total sugar, titratable acidity, pH, phenol, anthocyanin, lactic acid, ethanol and DPPH scavenging activity. The composition of must and herbal wine is shown in Table 7.7. Must in this study was the medicinal plant parts ameliorated saccharified purple sweet potato juice before the fermentation process.

Table 7.7 Biochemical composition of herbal purple sweet potato must and wine with standard deviation.

Parameter	Must (mean value)	Standard deviation	Wine (mean value)	Standard deviation
TSS(°Brix)	20.88	0.12	4.00	0.44
Reducing sugar (g/100 ml)	5.88	0.22	0.38	0.12
Starch (g/100 ml)	3.22	0.10	0.24	0.06
Total sugar	14.00	0.16	0.95	0.07
Titratable acidity (g tartaric acid/100 ml)	0.52	0.32	1.25	0.04
pH	5.32	0.29	3.34	0.12
Phenol (g/100 ml)	0.21	0.19	0.19	0.02
Anthocyanin (mg/100 ml)	68.21	0.08	59.90	1.30
Lactic acid (mg/100 ml)	0.02	0.33	1.92	0.32
Ethanol (% v/v)	0	0.23	8.61	1.88
DPPH (%)	70.13	0.26	51.35%	0.19

7.4.4.1 Acidity

The titratable acidity (TA) increased from 0.52 g tartaric acid/100ml in must to 1.25 g tartaric acid/100ml in wine. The increase of TA was concomitant with a decrease in pH. The pH of the must was 5.32 which was reduced in the herbal wine (pH, 3.34). The must had a lactic acid content of 0.02 mg/100 ml which increased to a value of 1.92 mg/100 ml in the wine. The acidic characteristics of the herbal wine was similar to that of the red wine (TA, 1.34 g/100 ml; pH, 3.61; lactic acid, 1.14 mg/100 ml) produced from purple sweet potato.

7.4.4.2 Carbohydrates

As expected the carbohydrate contents (starch, total sugar, reducing sugar and TSS) decreased in the course of the alcoholic fermentation process for wine production. The starch, total sugar and reducing sugar of the must were 3.22 (g/100 ml), 14.00 (g/100 ml) and 5.88 (g/100 ml), respectively. The TSS of the must was 20.88°Brix. The herbal purple sweet potato wine contained: 0.24 g/100 ml of starch, 0.95 g/100 ml of total sugar, 0.38 g/100 ml of reducing sugar and TSS of 4°Brix. The reduction of carbohydrate contents of must to wine was indicative of the utilization of sugars by the wine yeast for ethanol production.

7.4.4.3 Antioxidants

Phenol and anthocyanins were estimated in the must as well as in the herbal purple sweet potato wine. The must had a total phenol content (TPC) of 0.21 (g/100 ml) which was a little higher than that of the TPC of herbal purple sweet potato wine (0.19 g/100 ml). Total phenol content (TPC) of the herbal wine is actually equivalent to that of red wines (Sanchez–Moreno et al. 1998, Piaxao et al. 2007). TPC content of the herbal wine might be due to the cumulative polyphenol contents of purple sweet potato and herbal additives. The anthocyanin content of the herbal purple sweet potato wine was estimated to be 59.90 mg/100 ml which was lower than that of the must (68.21 mg/100 ml). The anthocyanin content of the herbal wine was higher than that of the red wine from purple sweet potato roots (55.09 mg/100 ml) as discussed in Section (7.3.5.3). Antioxidative properties could facilitate the proliferation of cells in the damaged parts of the body hence hastening the collagen synthesis. DPPH scavenging activity of the must and herbal purple sweet potato wine were 70.13% and 51.35%, respectively at a dose of 250 µg/ml.

7.4.4.4 Alcohol content

The wine (Fig. 7.4) was moderately alcoholic (8.61% v/v) which was lower than that of the red wine obtained from purple sweet potato roots (9.33% v/v) in our previous study.

Similar types of biochemical studies have been carried out on several herbal wines or wines having medicinal attributes. Herbal wine was prepared from three different clones of infused tea leaves which had a pH in the range of 4.10–4.50, titratable acidity (TA) in the range of 1.21–1.40 and ethanol content in a range of 4.4–6.0 (Aroyeun et al. 2005). The pH, TA and ethanol content of the tea wines were comparable to that of the herbal purple sweet potato wine. However the total phenolic content (TPC) of the herbal wine was higher than that of the tea wines. The poly–phenols in the tea wines after a period of

Figure 7.4. Herbal purple sweet potato (PSP) wine.

6-months maturation varied from 0.0013–0.0091 g/100 ml. Poly-phenols are known to operate as free radical scavengers quenching hydroxyl radicals (•OH) or superoxide anion radicals ($O•^{-2}$) (Sichel et al. 1991). Several studies conducted on epidemiology show that food and beverages rich in flavonoids possess cardioprotective properties and show favourable impacts in cardiogenesis and other cardiac diseases (Soleas et al. 1997). Additionally the low pH (3.34) of the herbal purple sweet potato wine is highly favourable for the stability of the poly-phenols as they auto-oxidize when the pH increases (Mochizuki et al. 2002).

In another study jamun (*Syzgium cumini* L.) fruits were fermented to wine. The authors claim that the wine contains medicinal properties such as anti-diabetic and anti-bleeding in case of patients with piles. The jamun wine had a TSS of 2.8°Brix, reducing sugar of 0.49 g/ 100 ml, TA of 1.1 g tartaric acid/100 ml, pH of 3.3, TPC of 0.22 g/ 100 ml, anthocyanin content of 60 mg/100 ml, tannin of 1.40 mg/ 100 ml, lactic acid of 0.80 mg/100 ml and ethanol concentration of 6% (v/v) (Chowdhury and Ray 2007). The biochemical studies of the jamun wine corroborated with the findings of herbal purple sweet potato wine. Indian gooseberry (*Emblica officinalis*) was fermented into wine. The biochemical studies of the wine during various maturation periods showed that the wine had a pH which varied from 3.30 to 3.52, phenolics varied in a range of 0.95–1.75 g/100 ml, TSS varied from 4.6–4.8%, sugar content varied from 0.50–0.60 g/100 ml and ethanol content in a range of 8.90–9.00% (v/v) (Soni et al. 2009). Although the TPC and ethanol content of the herbal purple sweet potato wine were lesser than that of the wine prepared from Indian gooseberry fruits, the other parameters (pH, TSS, sugar content) showed similar trends with that of the herbal purple sweet potato wine.

Alcoholic beverages have been produced from Black mulberry (*Morus nigra*), a fruit popular for both its nutritional attributes and flavour, and its conventional use in natural medicine as it is rich in bioactive compounds such as anthocyanins, phenols, potassium and citric acid contents. This fruit is used to prepare juices and alcoholic beverages since it is known to be therapeutically active for the control of Type II Diabetes mellitus as well as infections of the throat, tongue and mouth. The must obtained from the black mulberry was higher in anthocyanins (212 mg/100 ml) as compared to the must prepared for herbal wine in our study. The beverage had a pH of 3.47 and alcoholic grade of 7.40 similar to that of the herbal purple sweet potato wine.

7.4.5 *FTIR analysis*

Analysis of phytochemicals and chemical constituents of medicinal herbs with the use of the infrared (IR) spectroscopical method was initially limited only for structural elucidation of isolated compounds from the herbal matrices. Presently IR is also applicable in phytochemical studies as 'fingerprinting' devices, for comparison of natural with synthetic samples (Harborne 1998).

This method has been described in Section 7.3.6. Samples of herbal purple sweet potato wine and a standard commercial wine (Port wine 1000) were subjected for detection and comparison of functional groups in the procedure described earlier in Section 7.3.6. The FTIR analysis and interpretation of herbal purple sweet potato wine were done in a similar method as that was applied for the analysis of red wine from purple sweet potato.

The spectra of the herbal purple sweet potato wine have been displayed in Fig. 7.5. The strong phenol O–H stretch detected at 3591/cm was indicative of the presence of alcohol and phenol groups in both

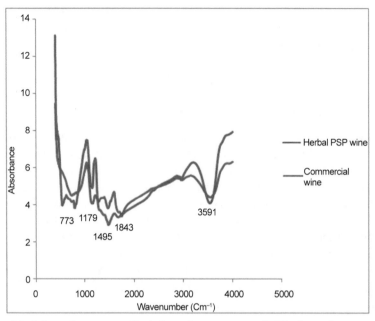

Figure 7.5. Infra–red spectrum of herbal purple sweet potato (PSP) wine and a commercial wine.

herbal and commercial wine. Similarly peaks detected at 1843/cm showed the presence of anhydrides group (C = O stretching) in both the samples. *Azo* compound (N = N stretching) was detected at 1495/cm, second overtone C–H stretching was detected at 1179/cm and alkynes (= C–H bending) was detected at 773/cm in both the wine samples. Further the C–N stretching bands showed the presence of amines. Aromatic amines showed bands in the range of 1323–1250/cm and aliphatic amines showed bands at 1220–1000/cm in both the samples. Primary (–NH$_2$), secondary (–NH) and tertiary (no hydrogen attached to N) amines may be differentiated by using the infrared spectra. Some additional new bands were detected in case of herbal purple sweet potato wine samples such as at C=O stretch bands at 1710/cm and aliphatic C–N stretching at 1269/cm. In another study, extension on the use of FTIR spectroscopy was done in conjugation with the appropriate chemometric methods (PCA and SIMCA) for the classification of *Orthosiphon stamineus* Benth, (known as Java tea).

The classification was based on geographical origin and varieties from acquired characteristics of infrared spectrum.

The study revealed that sample origin had a dominant effect on the chemical constituent of the plant rather than plant-to-plant varieties. The good classification model obtained from both PCA and SIMCA (Soft Independent Modeling of Class Analogy) further proved that classification of samples from various sourcing and varieties is possible with the incorporation of chemometric techniques and FTIR spectroscopy (Sim et al. 2004).

7.4.6 Medicinal attributes

Certain biological active compounds are present in the herbal additives such as nimbidin, margolone and margolonone in neem (*Azadirachta indica*) (Biswas et al. 2002), latex (activity against *Corynebacterium, Salmonella, Streptococcus, S. aureus*) in *Aloe vera* (Cowan 1999), allicin and ajoene in garlic (*Allium sativum*), atropine in belladonna (*Atropa belladona*), xanthotoxol, imperatorin, alloimperatorin and alkaloids like aegeline and marmeline in bael (*Aegle marmelos*) (Joy et al. 1998). Indian gooseberry (*Emblica officinalis*) is a rich source of natural vitamin C and contains cytokinin like substances namely zeatin, zeatin riboside and zeatin nucleotide (Joy et al. 1998). The basic cardamom (*Elleteria cardamomum*) aroma is produced by a combination of the major components, α-terpinyl acetate and 1, 8 – cineole. These aromatic compounds are reported to be antiseptic, anti-inflamatory, carminative and stimulating. Leaves of five leaved chaste (*Vitex negundo*) contain essential health beneficial compounds like alkaloid nishindine, flavonoids such as flavones, luteolin–7–glucoside, casticin, iridoid glycosides, carotenes, vitamin–C and C-glycoside (Husain et al. 1992). Eugenol (l–hydroxy–2–methoxy–4–allylbenzene), the bioactive constituent present in holy basil (*Ocimum sanctum* L.) is understood to be immensely responsible for the curative properties (Prakash and Gupta 2005). Roots of asparagus (*Asparagus adscendens*) contain several alkaloids such as asparagin; which has anticancer properties. It also contains a number of antioxytocic saponins, viz. Shatavarisn – I to IV (Syamala 1997). Medicinal food and beverages are supplemented in various nutrients and phytochemicals, which confer specific health benefits to it. These chemical compounds of the medicinal plants described above presumably present in the purple sweet potato herbal wine may contribute many health benefits to the consumer.

7.4.7 Statistical analysis

Correlation and principal component analysis were applied to the analytical variables of herbal purple sweet potato wine and the sensory scores of herbal purple sweet potato wine and a reference wine were subjected to independent '*t*' test as in case of red wine from purple sweet potato (Section 7.3.7)

7.4.7.1 Correlation of the analytical and proximate data of herbal purple sweet potato wine

The correlation analyses among the proximate and analytical variables are presented in Table 7.8. The most important correlations were starch–TSS (–0.524), ethanol–reducing sugar (–0.556), LA–anthocyanin (0.625), DPPH–anthocyanin (0.549), DPPH–LA (0.815).

The correlation between DPPH and anthocyanin pigments signifies that the free radical scavenging activity of the herbal wine is due to the presence of anthocyanin pigments. Similar results of correlation were obtained from the analytical variables of purple sweet potato wine.

7.4.7.2 Principal component analysis

Using PCA, the 11 original proximate and analytical variables were reduced to four principal components (PC1–PC4), which had eigen values larger than 1 and retained for rotation (Hair et al. 1998). PC1, PC2, PC3 and PC4 accounted for 24.18, 20.18, 15.41 and 14.76%, respectively and the total variations (74.53%) are presented in Table 7.9. To assist the interpretation of dimensions, the factor pattern was rotated using the varimax method. Based on the guidelines provided by Stevens (1992), an attribute was correlated to load heavily on a given component if the factor loading was greater than 0.72. All the 11 analytical variables loaded heavily on the dimension. Four analytical variables, i.e., total sugar (+ve), anthocyanin (+ve), DPPH (+ve) and

Table 7.8. Correlation coefficient for analytical + proximate variables of herbal purple sweet potato wine.

TSS	Reducing sugar	Starch	Total sugar	Titratable acidity	pH	Phenol	Anthocyanin	Lactic acid	Ethanol	DPPH
1.000	-0.112	-0.524	0.187	-.0293	-0.386	0.037	0.000	-0.341	0.333	-0.398
	1.000	0.048	0.071	0.256	-0.027	-0.048	-0.049	0.193	-0.556*	0.046
		1.000	0.214	0.209	0.273	-0.071	0.099	0.386	0.026	0.414
			1.000	0.163	-0.055	-0.357	0.247	0.410	0.159	0.184
				1.000	0.133	-0.116	0.241	0.094	0.026	0.090
					1.000	-0.055	0.255	0.248	-0.030	0.316
						1.000	-0.247	-0.337	0.000	-0.092
							1.000	0.625*	0.220	0.549*
								1.000	-0.107	0.815**
									1.000	-0.102
										1.000

*Correlation is significant at the 0.05 level (2-tailed).
** Correlation is significant at the 0.01 level (2-tailed).

121

Table 7.9. Principal components analysis of proximate variables (herbal purple sweet potato wine).

Parameter	PC 1	PC 2	PC 3	PC 4
TSS	0.158	–0.909	–0.137	–0.145
Reducing Sugar	0.118	–0.006	0.949	0.198
Starch	0.258	0.721	0.113	–0.036
Total sugar	0.727	–0.391	–0.295	0.176
Titratable acidity	0.080	0.057	0.324	0.827
pH	0.101	0.579	–0.052	0.227
Phenol	–0.156	–0.194	0.168	–0.685
Anthocyanin	0.747	0.096	0.008	0.373
Lactic acid	0.872	0.089	0.432	0.134
Ethanol	0.048	–0.436	–0.593	0.359
DPPH	0.831	0.443	0.033	–0.202
Variance explained (%)	24.18	20.18	15.41	14.76
Eigen value	3.28	2.26	1.38	1.33

Extraction Method: Principal Component Analysis.
Rotation Method: Varimax with Kaiser Normalization (Eigen value >1).

lactic acid (+ve) were loaded heavily on PC1. Total soluble solid (–ve) and starch (+ve) have been loaded substantially on PC2. Reducing sugar (+ve) and TA (+ve) were loaded on PC3 and PC4, respectively. Plotting PC1 vs. PC2, PC1 vs. PC3 and PC2 vs. PC3 loading, depicted these relationships visually in Fig. 7.6a, b and c, respectively.

7.5 Anthocyanin Rich Beer from Purple Sweet Potato

Beer is generally made by using malt prepared from barley (Bamforth 2004). As barley is not available in several regions it is either partially replaced by adjuncts such as rice or maize flakes or wholly replaced with by cereals such as sorghum and wheat (Owuama 1997, Eblinger 2009). Adjunct is defined as an additional additive which does not change the original characteristics of the final product. Beer preparation has been also successfully carried out by using fruits as adjuncts. Fruits such as banana and cherry have been utilized as a substrate for the preparation of fruit beer (Carvalho 2009, Bonciu and Stoicescu 2008). A study has also been conducted by taking sweet potato flour as an adjunct in the brewing of sorghum beer (Etim and Ektokakpan 1992).

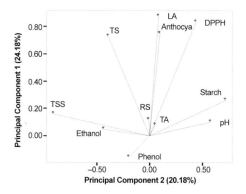

Figure 7.6a. Graphical representation of principal components PC1 vs. PC2 of analytical variables of herbal purple sweet potato wine.

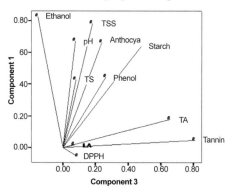

Figure 7.6b. Graphical representation of principal components PC1 vs. PC3 of analytical variables herbal purple sweet potato wine.

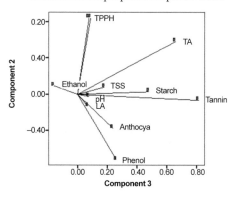

Figure 7.6c. Graphical representation of principal components PC2 vs. PC3 of analytical variables of herbal purple sweet potato wine.

In the usual conditions of beer preparation, malt (dried sprouted barley) is grinded to form a mass called grist. In the present study, grist (ground malt) was replaced by various concentrations (30%, 50% and 100%) of purple sweet potato flakes. The wort and beer prepared by using purple sweet potato flakes were subjected to biochemical analyses. Sensory evaluation was carried out to determine the acceptance of the beer samples prepared using purple sweet potato as the adjunct.

7.5.1 Saccharification

Saccharification of the substrates was carried out by using the enzyme promalt and incubating the samples at different temperatures, i.e., 42°C, 50°C, 62°C and 70°C in sequence. Promalt is commonly used for the production of alcohol from cereals such as wheat, maize, etc. (Agu et al. 2008) as well as beer from barley and sorghum (Bajamo and Young 1992, Goode et al. 2005).

The conversion rate of starch to fermentable sugars was different in the case of different samples. The sample with 100% grist (control) rendered a starch conversion rate of 17.85%. Among the purple sweet potato beer samples, the sample prepared with 100% purple sweet potato flakes showed the highest conversion rate (30%) followed by the sample containing 50% purple sweet potato flakes (28.68%) and 30% purple sweet potato flakes (22.5%). This might be due to the presence of higher concentration of native ($1\rightarrow3$, $1\rightarrow4$)–glucanase (limiting in cereals) in sweet potato (Etim and Ektokakpan 1992).

7.5.2 Stock culture and starter culture

The stock culture (*S. cerevisiae*) was the same used for the preparation of red wine and herbal wine (Section 7.3.2 and Section 7.4.2).

Starter culture was prepared by inoculating a loop full of stock culture with a MYGP medium and was further incubated at 20°C in a BOD incubator for 24 hrs.

7.5.3 Beer making process

The steps in beer making involve the primary processing of purple sweet potato, i.e., washing, de-skinning, slicing and sun-drying to prepare purple sweet potato flakes. After the primary processing the other steps are saccharification, blending of flakes with grist (ground malt), amelioration of wort with dextrose, addition of hops, fermentation, finishing and bottling. The entire process of beer making has been described in Fig. 5.3 of the Materials and Methods (Section 5.7.3).

7.5.4 Biochemical analyses

The biochemical compositions of the wort and the beer samples are given in Table 7.10 and 7.11, respectively. The parameters of interest were TSS, starch, total sugar, specific gravity, pH, titratable acidity, lactic acid, phenolic content, anthocyanin, ethanol and DPPH scavenging activity. The wort in this study was the ameliorated juice of the saccharified and boiled mash of purple sweet potato flakes and grist combinations, and the beer was the undistilled carbonated alcoholic beverage made out of the wort after fermentation with *S. cerevisiae*.

7.5.4.1 Acidity

7.5.4.1.1 Wort

The pH of the purple sweet potato wort samples varied from 5.30 to 5.40 [$F(3,8) = 22.61$]. The control (100% grist) wort sample had a pH of 5.35. The titratable acidity (TA) was 0.42, 0.35, 0.36 and 0.24 g tartaric acid/100 ml for the wort samples prepared with 0, 30, 50 and 100% purple sweet potato flakes as adjunct, respectively. The lactic acid content in the wort samples was negligible ranging from 0.02 to 0.08 mg/100 ml.

Table 7.10. Biochemical composition and one way ANOVA for the parameters of wort prepared from different concentrations of purple sweet potato flakes and control.

Parameters	Treatments				F–value	p
	100% grist (Control)	30% purple sweet potato + 70% grist	50% purple sweet potato +50% grist	100% purple sweet potato		
TSS (°Brix)	9.77 ± 0.34	9.93 ± 0.42	9.90 ± 0.23	9.97 ± 0.26	0.71	NS
Starch (g/100 ml)	1.81 ± 0.09[a]	1.80 ± 0.07[a]	2.00 ± 0.23[b]	2.90 ± 0.07[b]	162.74	**
Total sugar (g/100 ml)	23.02 ± 0.53[a]	26.99 ± 0.64[a]	28.14 ± 0.34[a]	15.10 ± 0.54[b]	15.46	**
Specific gravity	1.04 ± 0.22	1.03 ± 0.06	1.05 ± 0.03	1.04 ± 0.02	1.33	NS
pH	5.35 ± 0.04[a]	5.30 ± 0.08[b]	5.33 ± 0.09[a,b]	5.40 ± 0.54[c]	22.61	**
TA (g tartaric acid/100 ml)	0.42 ± 0.03[a]	0.35 ± 0.02[b]	0.36 ± 0.12[b]	0.24 ± 0.10[c]	225.00	**
LA (mg/100 ml)	0.02 ± 0.01[a]	0.05 ± 0.01[b]	0.02 ± 0.01[c]	0.08 ± 0.01[d]	280.19	**
Phenol (mg/ml)	0.19 ± 0.10[a]	0.20 ± 0.07[a]	0.30 ± 0.03[b]	0.42 ± 0.01[c]	415.60	**
Anthocyanin (mg/100 g)	0.00[a]	2.87 ± 0.21[b]	4.83 ± 0.12[c]	9.84 ± 0.15[d]	3027.98	**
Ethanol (%v/v)	0.00	0.00	0.00	0.00	0	-
DPPH (%)	1.20 ± 0.05[a]	3.57 ± 0.02[b]	5.43 ± 0.23[c]	8.63 ± 0.09[d]	17373.50	**

**Significant at 1% level.
NS Non-significant.
Different letters in the same row indicate significant differences (p < 0.05; Tukey's HSD).

Table 7.11. Biochemical composition and one way ANOVA for the parameters of beer prepared from different concentrations of purple sweet potato flakes and control.

Parameters	Treatments					
	100% grist (Control)	30% purple sweet potato + 70% grist	50% purple sweet potato + 50% grist	100% purple sweet potato	F –value	P
TSS (°Brix)	3.53 ± 0.58[a]	3.570.12[a]	3.0 ± 0.20[b]	3.00 ± 0.31[b]	18.18	**
Starch (g/100 ml)	0.61 ± 0.03[a]	0.77 ± 0.02[b]	0.63 ± 0.02[a]	0.47 ± 0.02[c]	111.29	**
Total sugar (g/100 ml)	6.15 ± 0.05	6.93 ± 0.04	6.93 ± 0.12	3.3 ± 0.40	201.92	**
Specific gravity	1.01 ± 0.01[a]	1.02 ± 0.01[b]	1.01 ± 0.01[b]	1.01 ± 0.01[a]	19.49	**
pH	3.05 ± 0.03[a]	3.05 ± 0.05[a]	3.40 ± 0.02[b]	3.45 ± 0.05[b]	97.12	**
TA (g tartaric acid/100 ml)	0.70 ± 0.02[a]	0.76 ± 0.03[b]	0.68 ± 0.02[a]	0.73 ± 0.02[a]	8.81	**
LA (mg/100 ml)	0.12 ± 0.01[a]	0.10 ± 0.01[a]	0.19 ± 0.02[b]	0.11 ± 0.01[a]	39.05	**
Phenol (mg/ml)	0.19 ± 0.02[a]	0.21 ± 0.02[a]	0.31 ± 0.01[b]	0.39 ± 0.03[b]	76.60	**
Anthocyanin (mg/100 g)	0 ± 0.00[a]	2.66 ± 0.57[b]	4.2 ± 0.10[c]	9.71 ± 0.05[d]	592.48	**
Ethanol (%v/v)	3.57 ± 0.06[a]	3.77 ± 0.06[b]	3.37 ± 0.06[c]	3.77 ± 0.06[b]	33.00	**
DPPH (%)	0.96 ± 0.01[a]	3.15 ± 0.02[b]	5.31 ± 0.07[c]	9.737 ± 0.02[d]	28514.18	**

Data are represented as ± SD, **Significant at $p < 0.01$ level, Different letters in the same row indicate significant differences ($p < 0.05$; Tukey HSD).

7.5.4.1.2 Beer

In contrast, the beer samples were more acidic [pH, 3.05–3.45; TA, 0.68–0.76 g tartaric acid/100 ml; LA, 0.10–0.19 mg/100 ml] than that of the wort samples. The pH was 3.05, 3.40 and 3.45 for the samples prepared with 30, 50 and 100% purple sweet potato flakes as the adjunct. The control beer sample (100% grist only) had a pH of 3.05. The results corroborate with the study by Palmer (2006) for malt beer in which the pH varied from 3 to 4.2. The TA was 0.76, 0.73, 0.68 and 0.70 g tartaric acid/100 ml for beer prepared with 0, 30 and 50 and 100% purple sweet potato flakes, respectively. Lactic acid content of the beer samples ranged from 0.10–0.19 mg/100 ml (Table 7.12).

Table 7.12. Principal component analysis of proximate variables (purple sweet potato beer).

Biochemical variables	Principal components		
	PC1	PC2	PC3
TSS	−0.880	0.344	0.089
Starch	−0.575	0.105	0.789
Sugar	−0.598	−0.384	0.691
Gravity	−0.036	−0.019	0.979
pH	0.932	−0.274	−0.140
TA	0.068	0.907	0.287
LA	0.235	−0.916	0.266
Phenol	0.971	0.000	−0.157
Anthocyanin	0.944	0.256	−0.180
Ethanol	0.063	0.958	−0.114
DPPH	0.964	0.171	−0.171
Variance explained (%)	46.92	27.52	21.25

Extraction Method: Principal component analysis.
Rotation Method: Varimax with Kaiser Normalization (Eigen value > 1).

7.5.4.2 Carbohydrates

7.5.4.2.1 Wort

The starch content was 1.81, 1.80, 2.00 and 2.90 g/100 ml for the wort samples prepared with 0, 30 and 50 and 100% purple sweet potato flakes, respectively. The total sugar contents were 23.02, 26.99, 28.14 and 15.10 g/100 ml for the samples prepared 0%, 30% and 50% and 100% purple sweet potato flakes, respectively. The TSS and specific gravity in the wort samples were adjusted to around 10°Brix and 1.040, respectively by the addition of dextrose. For this reason there were no significant differences in TSS [$F(3,8) = 0.71$] and specific gravity [$F(3,8) = 1.33$] in the wort samples.

7.5.4.2.2 Beer

As expected starch, total sugar, TSS and specific gravity fell significantly in the beer as compared to the wort samples.

The starch contents were 0.77, 0.63 and 0.47 g/100 ml for the beer samples prepared with 30, 50 and 100% purple sweet potato flakes, respectively. The control beer sample (100% grist) had a starch content of 0.61 g/100 ml. The beer with 30 and 50% purple sweet potato flakes as adjunct contained 6.93 g/100 ml of total sugar. The beer prepared with 0% purple sweet potato flakes and 100% purple sweet potato flakes as adjunct had a total sugar content of 6.15 and 3.3 g/100 ml, respectively. Significant differences were observed for TSS [$F_{(3,8)}$ = 18.18] and specific gravity [$F_{(3,8)}$ = 19.49] in the case of beer samples. The TSS fell to 3.53, 3.57, 3.0, 3.0°Brix in the beer samples prepared with 0, 30, 50 and 100% purple sweet potato flakes as adjunct, respectively.

7.5.4.3 Antioxidants

7.5.4.3.1 Wort

The wort prepared with 0, 30, 50 and 100% purple sweet potato flakes had 0, 2.87, 4.83 and 9.84mg/100ml of anthocyanin, respectively. The differences of anthocyanin content among the wort samples were highly significant [$F_{(3,8)}$ = 3027.98] . The wort prepared from 30% purple sweet potato flakes +70% grist, 50% purple sweet potato flakes + 50% grist and 100% of purple sweet potato flakes could be visually stinguished on the basis of the darkness of the pinkish colour, imparted by the anthocyanin pigments of the purple sweet potato (Fig. 7.7). The wort prepared with 0, 30, 50 and 100% purple sweet potato flakes contained 0.19, 0.20, 0.30, 0.42 mg/ml of phenol, respectively. The DPPH scavenging activity for wort samples prepared from 100% grist, 30% purple sweet potato flakes + 70% grist, 50% purple sweet potato flakes + 50% grist and 100% purple sweet potato flakes were 1.2, 3.6, 5.4 and 8.63%, respectively at a dose of 250 µg/ml.

7.5.4.3.1 Beer

As expected the concentration of anthocyanin in the beer raised along with an increase in the concentration of purple sweet potato

Figure 7.7. Fermentation of wort. wort prepared from (a) 30% purple sweet potato flakes + 70% grist, (b) 50% purple sweet potato flakes + 50% grist and (c) 100% purple sweet potato flakes.

Color version at the end of the book

flakes in the recipe of the wort preparation. Significant differences of anthocyanin content were observed in the cases of beer samples because of the variation in the concentration of purple sweet potato flakes (rich in anthocyanin) during the beer preparation [$F(3,8)$ = 592.48]. Beer samples prepared with 0, 30, 50 and 100% purple sweet potato flakes had anthocyanin content of 0, 2.66, 4.20 and 9.71 mg/100 ml, respectively. The anthocyanin concentration in the purple sweet potato beer was comparably lower than that of wines prepared from purple sweet potato (55.09 mg/100 ml) in our previous study as well as for wine made out of grapes (60–67 mg/100 ml), jamun fruits (60 mg/100 ml) (Chowdhury and Ray 2007) and alcoholic beverages obtained from black mulberry (210 mg/100 ml) (Darias–Martin et al. 2003). Anthocyanin is an important constituent of purple sweet potato as it is more stable than that of other fruits and vegetables like strawberry, red cabbage etc. Phytochemicals like anthocyanin are substances of plant origin those may be ingested by human beings daily in substantial quantities, which exhibit the potential for modulating metabolism so as to prevent cancer and cardiovascular diseases (Rincon–Leon 2003).

Similarly the beer prepared with 0, 30, 50 and 100% purple sweet potato flakes contained 0.19, 0.21, 0.31 and 0.39 mg/ml of phenols, respectively [$F_{(3,8)} = 76.60$]. The purple sweet potato beer had 3.15, 5.31 and 9.73% of DPPH scavenging activity for the same samples (30, 50 and 100% purple sweet potato flakes as adjunct) in chronological order. Significant differences in DPPH scavenging activity was observed among the beer samples prepared with different concentrations of purple sweet potato flakes [$F_{(3,8)} = 28514.18$].

Anthocyanin and phenols are considered as antioxidants. The data signifies that the phenol concentration increased with the increase of purple sweet potato concentration in the wort and beer. According to some researchers, levels of antioxidants in the beer are in comparable magnitude to fruit juices, teas and wines (Joe et al. 1999, Gorinstein et al. 2000). Flavonoids in foodstuffs have attracted the most attention for their potential value as chemoprotective agents (Horvathova et al. 2001). The polyphenols derived from beer are much more efficacious as inhibitors of the oxidation of LDL (low density lipo–protein) and CLDL (carbamylated low density lipo–protein) lipoproteins than ascorbic acid, alpha-tocopherol and beta-carotene (Vinson et al. 1995). Further the chemistry of the essential oil fraction of hops is enormously complex. From our study, it is envisaged that phenolic contents of purple sweet potato beer was attributed from the cumulative contribution of phenols from the grist, hops and purple sweet potato (Verzele 1986).

The DPPH scavenging activity declined in the beer during the process of fermentation.

7.5.4.4 Alcohol content

The beer samples prepared after finishing (Fig. 7.8) was slightly alcoholic. Ethanol content of the beer samples varied from 3.37–3.77% (v/v).

Indigenous beer is produced in different localities in different names. The prime raw material and the adjunct for the production depended upon the availability of the substrates or raw materials. Beers in Africa are known by different traditional names such as *'buru-kutu'*, *'otika'*, and *'pito'* in Nigeria, *'Nandis'* in Kenya, *'mowa'* in Malawi, *'kaffir beer'* in South Africa, *'merisa'* in Sudan, *'bouza'* in Ethiopia and *'pombe'*

Figure 7.8. Beer samples. Beer prepared from (a) 30% purple sweet potato flakes + 70% grist (b) 50% purple sweet potato flakes + 50% grist, (c)100% purple sweet potato flakes and from (d) 100% malt.

Color version at the end of the book

in many parts of East Africa. Sorghum is often mixed with maize (*Zea mays*) or millets, (*Pennisetum* spp.). In some cases such as in Central Africa, e.g., Zimbabwe, maize may form the major cereal component. Outside Africa sorghum is not used normally for brewing except in the U.S. where it is occasionally used as an adjunct. The method for producing these sorghum beers of the African continent as well as their natures is remarkably similar to the beer produced using purple sweet potato flakes in the present study. The major features of the traditional African beers comparable to the purple sweet potato beer are pinkish in colour, consumed without ageing and without the removal of microorganisms. '*Buru-kutu*' and '*Pito*' are two similar types of beer. But in the preparation of '*Pito*' adjuncts are not added whereas in the preparation of '*Buru-kutu*' adjuncts such as tropical tuber crop derivatives namely cassava flour (*gari*) is added. '*Bouza*' is an Ethiopian alcoholic beverage prepared from both wheat and maize (Okafor 2006). The beverage had a pH of 3.6 and alcohol content of 4% which was analogous to that of the characteristic of the beer produced from purple sweet potato flakes.

Koedo Brewery of Japan has been successfully using roasted sweet potato as a raw material for the manufacture of beer since 1996. The beer has moderate ethanol content with slight sweetness in its taste.

The alcohol content of the beers obtained from purple sweet potato flakes is some what lesser [3.37–3.77% (v/v)] than that of the commercial beer of Japan. *Shochu* is a Japanese distilled alcoholic beverage indigenously prepared in Japan from sweet potato. The beverage contains a higher amount of ethanol (25–40%) than that of the beer produced from purple sweet potato flakes and other similar types of beverages. In Japan fermented alcoholic coloured beverages (wine and beer) like purple sweet potato beer are manufactured and marketed in the Kyushu Province (Yamakawa 1997, 2000). Several studies have been conducted by incorporating sweet potato in the preparation of alcoholic beverages which has proved certain advantages. The preparation of the beverages was similar to that of the beer produced from purple sweet potato flakes. Substitution of sorghum malt with sweet potato flour by 20% gave a higher amylase activity and better organoleptic quality of the beer (Etim and Ektokakpan 1992). The authors suggested the application of sweet potato in place of the extraneous enzymes in the preparation of sorghum beer to reduce the production cost.

7.5.5 *FTIR analysis*

Fourier Transform Infrared (FTIR) spectroscopy in combination with multivariate data analysis has been introduced for the quality control and authenticity assessment of spirit drinks and beer in official food control programmes (Lachenmeier 2007). Samples of purple sweet potato beer were subjected to FTIR analysis for detection and comparison of functional groups (Figs. 7.9, 7.10 and 7.11). The procedure of analysis was similar to that described for purple sweet potato and herbal wine. The spectra (0–5000/cm) were interpreted by using the guidelines of Stuart (2004).

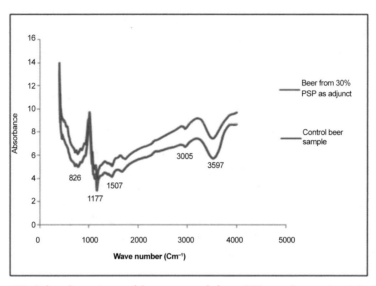

Figure 7.9. Infrared spectrum of beer prepared from 30% purple sweet potato (PSP) flakes + 70% grist and control (100% grist).

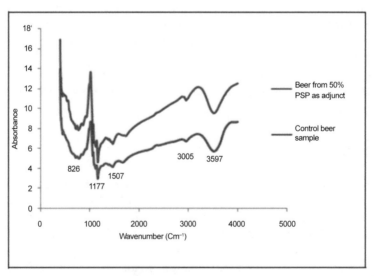

Figure 7.10. Infra–red spectrum of beer prepared from 50% purple sweet potato (PSP) flakes + 50% grist and control (100% grist).

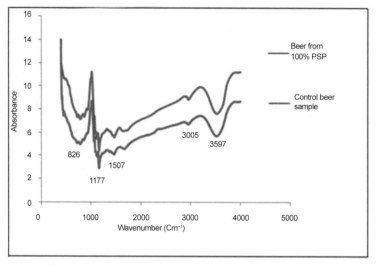

Figure 7.11. Infra–red spectrum of beer prepared from 100% purple sweet potato (PSP) flakes and control (100% grist).

All the beer samples showed the common peaks but the variations were observed only in the intensity of the peak. The peaks of the beer samples detected at 3597/cm are indicative of the presence of alcohol and phenol groups. However, the peaks (3597/cm) became sharper (Figs. 7.9, 7.10 and 7.11) in the samples containing a higher concentration of purple sweet potato flakes which confirmed the presence of a higher phenolic content. Minute peak was observed at 3005/cm in all the beer samples which represented the carboxylic acids. Carboxylic acids showed a strong broad O–H stretching band in the 3300–2500/cm range. Aromatic nitro compound NO_2 asymmetric stretching was detected at 1507/cm in all the beer samples. Aliphatic C–N stretching was observed in all the beer samples at 1177/cm. Similarly the peak detected at 826/cm showed NH_2 wagging and twisting in all the samples. Primary (–NH_2), secondary (–NH) and tertiary (no hydrogen attached to N) amines may be differentiated by using infrared spectra. All the amines showed C–N stretching bands, with aromatic amines showed bands in the 1360–1250/cm range and aliphatic amines showed bands at 1220–1020/cm.

7.5.6 *Statistical analysis*

One way ANOVA was applied to the biochemical parameters of the purple sweet potato wort and beer samples to find out the mean differences of the variants. ANOVA was also applied to the results of the sensory attributes obtained for the different beer samples to study the mean differences. Principal component analysis (PCA) was applied to the biochemical parameters of the beer samples to reveal the correlation between the parameters.

7.5.6.1 *ANOVA*

The one–way ANOVA of the major biochemical data of the wort samples varied significantly. Significant variations were observed for parameters like starch [$F(3,8) = 162.74$, $p = 0.000$], total sugars [$F(3,8) = 15.46$, $p = 0.001$], pH [$F(3,8) = 22.61$, $p = 0.000$], TA [$F(3,8) = 225.00$, $p = 0.000$)], LA [$F(3,8) = 280.19$, $p = 0.000$], phenol [$F(3,8) = 415.60$, $p=0.000$], anthocyanin [$F(3,8) = 3027.98$, $p = 0.000$] and DPPH activity [$F(3,8)= 17373.50$, $p = 0.000$]. *Post–hoc* analysis showed the significant difference among the treatments in starch, total sugars, reducing sugar, pH, TA, LA, phenol, anthocyanin and DPPH ($p < 0.05$, Tukey HSD). However, there was no significant difference of TSS and specific gravity among the wort samples. The ANOVA results of the biochemical parameters of wort samples of different variants are presented in Table 7.10.

Similarly biochemical parameters of beer samples were interpreted by using ANOVA. Significant differences observed were: TSS [$F (3, 8) = 18.18$, $p = 0.000$)], starch [$F (3, 8) = 111.29$, $p = 0.000$)], sugar [$F (3, 8) = 201.92$, $p = 0.000$)], specific gravity [$F (3, 8) = 19.49$, $p = 0.000$)], pH [$F (3, 8) = 97.12$, $p = 0.000$)], TA [$F (3, 8) = 8.81$, $p = 0.000$)], LA [$F (3, 8) = 39.05$, $p = 0.000$)], phenol [$F (3, 8) = 76.60$, $p = 0.000$)], anthocyanin [$F (3, 8) = 592.48$, $p = 0.000$)], ethanol [$F (3, 8) = 33.00$, $p = 0.000$)], and DPPH activity [$F (3, 8) = 28514.18$, $p = 0.000$)] among the beer samples.

The *post–hoc* analyses showed that the beer prepared with 100% grist (control) and with 30% purple sweet potato flakes as adjuncts were similar. The beer prepared with 50% of purple sweet potato flakes as adjunct was similar to that of the beer prepared from 100% purple sweet potato flakes. All the beer samples differed significantly in anthocyanin, ethanol and DPPH contents (p = 0.000; Tukey's HSD). The ANOVA results of the biochemical parameters of beer samples of different variants are presented in Table 7.11.

7.5.6.2 Principal component analysis

The biochemical data of the beer were subjected to principal component analysis (PCA) and the rotated component matrix is displayed in Table 7.12. Using PCA, the 11 original variables were reduced to three principal components (PC1, PC2 and PC3) which had eigen values larger than one and retained for rotation. PC1 accounted for 46.92% whereas PC2 and PC3 accounted for 27.52 and 21.25% of the total variations, respectively. When combined, PC1–PC3 together accounted for 95.69% of the total variations. All the 11 analytical variables were loaded heavily on three dimensions (Table 7.12). Five analytical variables, i.e., TSS (–), pH (+), phenol (+), anthocyanin (+) and DPPH (+) were loaded heavily on PC1. The correlation between Phenol (+), anthocyanin (+) and DPPH (+) indicated strong correlations among these attributes. Substantial factor loading of TA (+), LA (–) and ethanol (+) on PC2 and starch (+) and specific gravity (+) on PC3 was observed. The graphical representations of PC1 vs PC2, PC2 vs PC3 and PC1 vs PC3 are displayed in Figs. 7.12a, 7.12b and 7.12c, respectively.

Similar results were found for PCA projection of the physical and chemical parameters of 30 beer samples that accounted for 54.4 and 16.2% of the variability in the data on PC1 and PC2, respectively. The third PC accounted for only 7.9% of the variability (Langstaff et al. 1991).

7.5.7 Sensory evaluation of beer

The one way ANOVA of sensory data showed that all the samples differed significantly in terms of colour [F (4, 20) = 5.48, p = 0.000)],

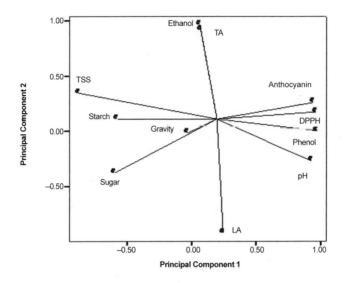

Figure 7.12a. Graphical representation of principal components PC1 vs. PC2 of analytical variables of purple sweet potato beer.

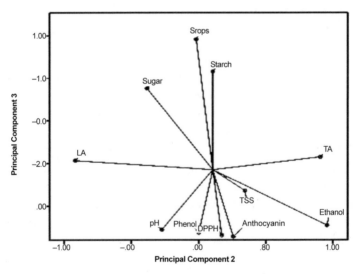

Figure 7.12b. Graphical representation of principal components PC2 vs. PC3 of analytical variables of purple sweet potato beer.

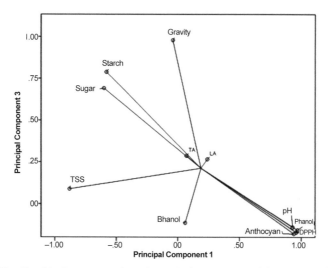

Figure 7.12c. Graphical representation of principal components PC1 vs. PC3 of analytical variables of purple sweet potato beer.

flavour [F(4,20) = 6.72, *p* = 0.000)] and acceptability [F(4,20) = 6.33, *p* = 0.000)] (Table 7.13). However, the samples scored similarly in clarity and foam. *Post–hoc* analyses showed that the beer prepared with 30% purple sweet potato flakes as adjunct differed significantly in colour and flavour (*p* = 0.001, Tukey's HSD) from others.

Similar studies were carried out by Donadini et al. (2008). Eighteen commercial beer samples were subjected to sensory evaluation by 51 consumers (26 male, 25 female). ANOVA results of descriptive analysis showed significant differences (p < 0.05) among the beers selected for the study for all the considered attributes. *F* statistics showed that beers can be significantly differentiated (p < 0.001) on the basis of their alcohol content (F = 58.484; p = 0.000), aroma (F = 53.711; p = 0.000), bitterness (F = 49.383; p = 0.000), body (F = 42.036; p = 0.000), flavour persistence (F = 38.934; p = 0.000), astringency (F = 24.816; p = 0.000), sweetness (F = 20.299; p = 0.000), acidity (F = 13.536; p = 0.000), degree of carbonation (F = 11.021; p = 0.000), and softness (F = 10.558; p = 0.000).

Table 7.13. One way ANOVA of sensory data of different treatments of beer.

Particulars	Treatments					F-value	P
	0% purple sweet potato flakes + 100% grist	30% purple sweet potato flakes + 70% grist	50% purple sweet potato flakes + 50% grist	100% purple sweet potato flakes + 0% grist	Commercial Beer		
Colour	7.20 ± 0.84^a	8.80 ± 0.45^b	6.60 ± 0.89^a	7.20 ± 0.84^a	7.20 ± 0.84^a	5.48	**
Clarity	7.00 ± 1.42^a	7.20 ± 0.84^a	6.80 ± 0.84^a	7.40 ± 1.14^a	7.20 ± 1.10^a	0.22	NS
Foam	6.60 ± 2.30^a	7.20 ± 0.84^a	6.80 ± 1.30^a	6.80 ± 0.84^a	6.40 ± 2.30^a	0.16	NS
Flavour	6.80 ± 1.09^a	8.80 ± 0.44^b	6.40 ± 0.5^a	7.00 ± 0.71^a	7.00 ± 1.00^a	6.72	**
Acceptability	6.00 ± 0.00^a	8.40 ± 0.55^b	7.20 ± 0.84^b	6.40 ± 0.55^a	6.40 ± 1.52	6.33	**

Data are represented as ± SD.
**Significant at p < 0.01 level.
NS – Not significant.
Different letters in the same row indicate significant differences (p < 0.05; Tukey HSD).

Figure 7.13. Trained panellists evaluating the purple sweet potato beer samples.

In our study, the overall acceptability of the samples was examined by the sensory panellists (Fig. 7.13) on Hedonic scores varying from 1–9. The panelists preferred the beer prepared with 30% purple sweet potato flakes as the adjunct based on the sensory parameters. The beer after proper bottling is shown in Fig. 7.14.

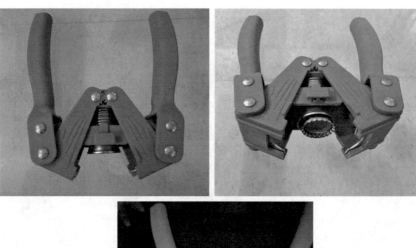

Figure 7.14. Crowning of beer after addition of priming sugar.

Results and Discussion

Technoeconomical and Cost Economics Report for Industrial Scale Production

◇◇

8.1 Technical Specifications of the Equipment Required

For production of purple sweet potato wine and purple sweet potato beer the equipment needed have been estimated considering the typical requirements of wineries and breweries along with the modifications necessary for the integrated purple sweet potato wine and beer manufacturing technology. The specifications are presented in Table 8.1 and Table 8.2.

8.2 Quality Control

The quality control section is presumed to have chemists, technicians and other personnel to assure that specific codes and standards of the industries are fulfilled. This section would have typical parameters

Table 8.1. Technical specifications for purple sweet potato wine and purple sweet potato beer manufacturing equipments.

Feed handling and conditioning		
Items	**Technical specification**	**Quantity**
Weighing scale	Weighbridge, 2 ton, with weight indication and recording	1
Unloading/moving fork lift	Capacity 0.5 ton, (diesel)	6
Sweet potato and barley storage unit/slab	1 m above ground level, slab of reinforced concrete, shed of epoxy coated steel, area 200 m²	6
Sweet potato (to peeling section) conveyor	Capacity 450 kg/hr, belt conveyor width 0.25 m, length 20 m, variable speed, 7.5 kW	4
Sweet potato and barley (to mashing section) conveyor	Belt conveyor, 450 kg/hr, width 0.25 m, length 10 m variable speed 5.5 kW	4
Mashing feeding conveyor	Belt conveyor, 450 kg/hr, width 0.25 m, length 10 m variable speed 4 kW	4
Magnetic separator (separation of impurities)	Belt type variable speed 5.5 kW	4
Shredder (for mashing)	Knife type, 45 kg/hr product size (to 62 mm), body of epoxy coated steel, knives of special alloy steel, variable speed, 75 kW	4
Mashed potato and crushed malt slurry conveyor	Screw conveyor, coated steel 450 kg/hr, variable speed, 7.5 kW	4
Washing screw conveyor	Screw conveyor, capacity 450 kg/hr with bottom strainers (mesh50), SS316, variable speed, 11 kW	4
Wash water tank	Steel, epoxy coated, 200 m³ capacity	2
Wash water pump	Centrifugal, cast iron, 50 m³/hr, 20 m head, 4 kW	4
Effluent wash water pump	Centrifugal submersible non-clogging, CI, 50 m³/hr, 20 m head, 4 kW	4
Clarifier/thickener	Steel, epoxy coated 100 m³ capacity with sludge scraper (0.3 kW)	4
Sludge pump	Progressive cavity type, 10 m³/hr, 50 m head, 3 kW	4
Belt filter press	Capacity 15 m³/hr, cake moisture content 70%	4
Clear water pumps	Centrifugal, 50 m³/hr, 20 m head, 4 kW	4
Filter cake conveyor	Screw conveyer, SS316, variable speed capacity 450 kg /hr, 0.2 m diameter, length 6 m, 1.5 kW	4
Slurrying tank	SS316 tank with variable speed mixer of SS316, 11 kW, 20 m³ capacity	4
Slurry pump	Progressive cavity type, 50 m³/hr 20 m head, variable speed, 11 kW	4
Wash water tank	Steel epoxy coated 100 m³ capacity	4
Wash water pump	Centrifugal, CI, 50 m³/hr, 30 m head, 7.5 kW	4

144

Table 8.2. Equipments for production plant: purple sweet potato beer and wine.

Enzymatic saccharification unit		
Items	**Technical specification**	**Quantity**
Slurrying tank	SS316 tank, 10 m³ capacity, with SS316 mixer, variable speed, 11 KW	2
Slurry pump	Progressive cavity type, 50 m³/hr, 20 m head variable speed, 11 kW	6
Enzymatic saccharification reactor (wine) and enzymatic incubation reactor (EI) (beer)	Cylindrical, closed, dish end bottom, 2.5 m diameter, 5 m length, steel internally cladded with SS 316, equipped with 6 agitators with external variable speed motor, each 5.5 kW, with internal heating (peristaltic)	2+2
Boiling kettle	Cylindrical, closed, dish end bottom, 2.5 m diameter, 5 m length, steel internally cladded with SS 316	1
Yeast propagation vessel	Jacketed vessels of 5, 10, 20, 40 m³ capacity with mixers/agitators 11, 16, 30 kW with internal pumps and auxiliaries 18 kW	2
ES and EI reactor transfer pump	Centrifugal, non-clogging, SS316, 50 m³/hr, 10 m head, 7.5 kW	2+2
Beer fermentation tank	304 SS with chromium-nickel cylindrical, closed, dish end bottom, 2.5 m diameter, 5 m length	6
Wine fermentation tank	304 SS with chromium-nickel cylindrical, closed, dish end bottom, 2.5 m diameter, 5 m length	6

for wineries and breweries such as acidity, density, ethanol, reducing sugars, etc. The chemicals and instruments required for quality control section is depicted in Table 8.3.

8.3 Economic Aspects

Based on the above contributing factors, the technical and economic feasibility of the whole process has been estimated. For accuracy of the data, assistance of different online as well as trustworthy sources such as industry and financial experts have been consulted. With their aid, the data and figures provided is almost realistic.

Table 8.3. Quality control requirements.

Chemical/reagent/ consumables	Quantity required per annum	Unit	Annual cost	
			INR	USD
$CuSO_4$	11680	mg	7718.14	107.196
NaOH	12775	mg	111014	1541.861
H_2SO_4	4380	ml	332048	4611.778
Ethanol	1825	ml	299756	4163.278
HCl	1825	ml	112547	1563.153
Anthrone reagent	0.073	mg	1	0.014
$NaHSO_4$	9125	mg	71357.5	991.076
HNO_3	730	ml	26827.5	372.604
$HClO_4$	730	ml	20775.8	288.553
2,6-dichloro Indophenol reagent	730	mg	3613.5	50.188
Oxalic acid	730	mg	14	0.194
Folin-Ciocalteu reagent	365	ml	8807.45	122.326
Na_2CO_3	9125	g	119701	1662.514
Saccharose	9.125	mg	2172.6	30.175
Acetic Acid	54750	ml	31930	443.472
$Ca(OH)_2$	365	g	722.7	10.038
Potassium dihydrogen phosphate	496.4	g	3372.8	46.844
Disodium hydrogen phosphate	3047.75	g	68094	945.750
Sodium hydrogen carbonate	4562.5	g	68094	945.750
Cinchonine base	547.5	g	28603	397.264
2,2-diphenyl-1picryl hydrazyl	730	mg	36879	512.208
Butylatedhydroxytoluene	82125	mg	5310	73.750
Petridishes			487494	6770.750
Borosilicate test tubes			50000	694.444

Table 8.3 contd. ...

... Table 8.3 contd.

Instruments required				
Instrument	Quantity	Price/ unit (INR)	Total	
			INR	USD
Weighing machine	4	100000	400000	5555.556
Kjeldahl flask	2	30000	60000	833.333
ASTM distillation unit	4	10000	40000	555.556
Water bath	10	5000	50000	694.444
Centrifuge	5	15000	75000	1041.667
Spoutless beakers	20	500	10000	138.889
Hot plate	5	5000	25000	347.222
Buchner funnel	5	500	2500	34.722
Silica basins	5	5000	25000	347.222
Oven	5	10000	50000	694.444
Desiccator	5	1000	5000	69.444
Muffle furnace	4	20000	80000	1111.111
Pestle mortar	5	500	2500	34.722
Refrigerated centrifuge	2	20000	40000	555.556
UV-Vis Spectrophotometer	1	600000	600000	8333.333
pH meter	2	7000	14000	194.444
Hand refractometer	2	1000	2000	27.778

8.3.1 Basis of estimation

The assumptions of cost projection, profit statement and capacity are as follows:

- The production of purple sweet potato wine per day is 2500 litres and that of purple sweet potato beer is 7500 litres.
- Salary/wages per month (INR) of the personnel is as follows: 1 Managing Director @ 3 lakh (4167 USD) + 3 Directors (Production

147

and quality; marketing and promotion; finance) @ 2 lakh (2777 USD), + 2 General Managers (Marketing and Production) @1 lakh (1388 USD) + 2 Assistant General Managers @ .75 lakh (1041 USD) + 4 Engineers @ 0.35 lakh (486 USD) + 3 Markting Officers @ 0.5 lakh (695 USD), 1 Finance Officer @ 0.5 lakh (695 USD) + 2 Accountants @ 0.25 lakh (347 USD) + 20 Technicians @ 0.15 lakh (208 USD) + 40 Labours @ 0.1 lakh (139 USD) + 2 Microbiologists @ .25 lakh (347 USD)+ 2 Chemists @ 0.25 lakh (347 USD).

- The Maximum Retail Price (MRP) of purple sweet potato wine is Rs. 320/– per litre (4.44 USD), government taxes cum duties and retailer's margin is 35% and 20% respectively. Hence the dealer's price is Rs. 166/– per litre (2.30 USD).
- Similarly, the MRP of purple sweet potato beer is Rs. 120/– per litre (1.66 USD), government taxes cum duties is 35% and retailer's margin is 20%. So the dealer's price is Rs. 62.4/– per litre (0.86 USD).

The cost of purchased equipment (PE) was available from Indian standard manufacturers. Moreover, superfluous costs like piping, civil works, land cost, etc., have been considered according to the Indian scenario. For example, the cost of land required for plant setup has based on the information obtained from Odisha Industrial Infrastructure Development Corporation (IDCO) (http://www.idco.in/2009/landrate.aspx). Other information such as engineering and supervision, consultancy fees, promotion and marketing has been obtained from typical liquid solid plant owners locally. The parameters used for estimating total cost of the project are thus almost realistic. Therefore, the deviations in calculation due to ever-changing market parameters may not be higher or lower than five percent. The total cost of the project on this basis is mentioned in Table 8.4. Individual expenses are depicted in the form of bar chart in Fig. 8.1 while Fig. 8.2 illustrates the percentage of all the individual costs which adds up to the total cost of project.

Based on the data obtained from the Indian market scenario, the estimated Annual Operation Cost (AOC) or Annual working capital

Table 8.4. Cost of the project.

	Nature of expenses	Total amount (lakhs INR)	Total amount (million USD)
A	Capital expenditure		
	Purchased equipment (PE)	136.180	0.189
	Equipment installation	40.854	0.057
	Piping	27.236	0.038
	Civil works	34.045	0.047
	Electrical works	13.618	0.019
	Instrumentation and control	37.450	0.052
	Land	75.000	0.104
	Others	40.854	0.057
	Total capital expenditure (A)	**405.237**	**0.563**
B	Revenue expenditure		
	Engineering and supervision	32.419	0.045
	Consultancy fees	32.419	0.045
	Contingencies	40.524	0.056
	Promotion and marketing	1000.000	1.389
	Total revenue expenditure (B)	**1105.361**	**1.535**
	Total investment cost (A+B)	**1510.598**	**2.098**

* The basis of cost estimation in the table (Table 8.4) is: equipment installation = 30% of PE; piping = 20% of PE; civil works = 25% of PE; electrical works = 10% of PE; instrumentation and control = 20% of PE; others (miscellaneous) = 30% of PE and land cost = as per the IDCO website (Tewfik et al. 2015, http://www.idco.in/2009/landrate.aspx).

requirement for production of purple sweet potato wine and purple sweet potato beer is briefed in Table 8.5. The annual operation cost includes chemicals required for winery and brewery, the salary of production, quality control, marketing personnel, store costs, utilities, and maintenance, which adds up to the total operation costs. The salary breakup of personnel is also tabulated in Table 8.6. This cost of production is projected till the year 2027 for getting a better picture of the scenario. Moreover, the total cost per litre of production of purple sweet potato wine and purple sweet potato beer is tabulated in

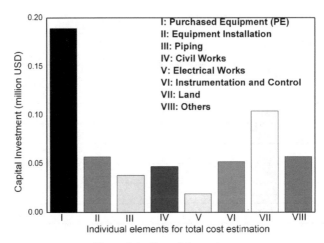

Figure 8.1. Cost of the project.

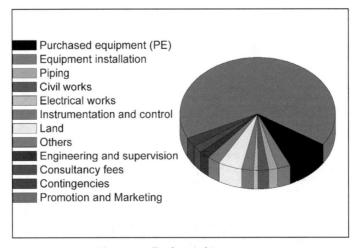

Figure 8.2. Total capital investment.

Color version at the end of the book

Table 8.7. Calculation of total cost of production of purple sweet potato wine and beer enabled us to conveniently project the net production for each year. The sale values of wine and beer has been estimated from the data obtained for production and plant operation cost, and is presented

Table 8.5. Annual Working capital requirement.

Particulars	Amount	
	Lakh INR	**Million USD**
Chemicals	18.788	0.026
Salary	292.800	0.407
Buffer Storage of finished products (20% of Annual Production cost)	186.374	0.259
Sundry Debtors	25.000	0.035
Stores & Spares	2.000	0.003
Maintainance	8.105	0.011
Utilities (Electricity, diesel, water)	80.000	0.111
Total	613.067	0.851
Interest on Cash Credit (35% of Total Working Capital)	27.895	0.039
Interest on Term Loan	120.848	0.168
GRAND TOTAL	**2272.407**	**3.156**
MEANS OF FINANCE		
Promoter's Contribution (60% of Total Investment)	906.359	1.259
Term loan (40% of Total Investment)	604.239	0.839
Cash Credit (35% of Annual Working Capital)	214.573	0.298
TOTAL CAPTAL REQUIRED	**1725.171**	**2.396**

in Table 8.8. An assumption was made that total capacity utilization for the year 2020 would be 50% of the total production volume. The percentage of production would go up by 10% every year till it reaches the year 2025 where it has been considered that the production would be 100% of the capacity. Keeping the above facts in mind the table has been formulated. On the same basis, yearly projection of profit and sales statement is presented in Table 8.9. A graphical representation of the annual profit and the amount of premium to be paid to the bank is

Table 8.6. Man power cost.

Particulars	Monthly salary (in lakhs INR)	Annual salary	
		Lakhs (INR)	Million USD
Director-factory	2.00	24.00	0.033
GM & AGM - production	1.60	19.20	0.027
Engineers	1.20	14.40	0.020
Technicians	4.00	48.00	0.067
Labourers	4.00	48.00	0.067
Total Factory MOH (A)	**12.80**	**153.60**	**0.213**
Total Factory MOH (Per Unit)		**4.21**	**0.006**
MD	3.00	36.00	0.050
Director-marketing	2.00	24.00	0.033
Director-finance	2.00	24.00	0.033
GM & AGM-marketing	1.60	19.20	0.027
Marketing officer	2.00	24.00	0.033
Finance officer	0.50	6.00	0.008
Accountant	0.50	6.00	0.008
Total Admin MOH (B)	**11.60**	**139.20**	**0.193**
Total Admin MOH (per Unit)	–	**3.81**	**0.005**
Total (A+B)	**24.40**	**292.80**	**0.407**
Total MOH per unit	–	**8.02**	**0.011**

also provided in Fig. 8.3. As evident from the figure, it is estimated that the premium becomes zero during the start of financial year of 2025 and a steady plateau would also be reached in terms of profit for both purple sweet potato wine and purple sweet potato beer. Moreover, the sales versus profit for winery and brewery are also depicted in Fig. 8.4. It is clear from the diagram that sales and profit goes hand in hand and from the year 2025 both reach a steady plateau which indicates the proposed project would be a stable business.

Table 8.7. Calculation of manufacturing cost per unit of purple sweet potato wine and purple sweet potato beer.

Wine			Beer		
Particulars	Qty in gram	Cost(in INR)	Particulars	Qty in gram	Cost in INR
Raw materials			**Raw materials**		
Purple sweet potato roots	600.00	7.50	Sweet potato flakes	30.00	0.60
Termamyl	1.00	0.01	Barley	120.00	1.20
Dextrozyme	5.00	0.14	Promalt	0.10	0.02
Ammonium sulfate	0.10	0.92	Hop pellets	0.08	0.50
Sodium metabisulfite	0.10	0.40	Dextrose	40.00	1.60
			MYGP	1.00	6.00
Direct material cost		**8.98**			**9.92**
Direct labour cost		**4.21**			**4.21**
Total Direct Cost (A)		**13.19**			**14.12**
Indirect Cost					
Chemicals		0.51			0.51
Maintainance		0.22			0.22
Finishing materials (bottles, cartons, crowns, corks, cans, labels, etc.)		7.50			7.50
Total Indirect Cost (B)		**8.24**			**8.24**
Total Cost Of production (A+B)		**21.43**			**22.36**

Return on Capital Employed (ROCE), a study of profitability ratio, measures how efficiently a company is using its capital. Simply put, ROCE measures how well a company is using its capital to generate profits. The ROCE is considered as one of the best profitability ratios and is commonly used by investors to determine whether a company is suitable to invest in or not. Higher ROCE is good for the company and preferable for investment.

Table 8.8. Schedule of production and sale of purple sweet potato wine and purple sweet potato beer.

Year ended 31st March	2020	2021	2022	2023	2024	2025	2026	2027
	Projected	Projected	Projected	Projected	Projected	Projected	Projected	Projected
Capacity utilisation	50%	60%	70%	80%	90%	100%	100%	100%
Production (wine)	0.456	0.547	0.638	0.730	0.821	0.912	0.912	0.912
Production (beer)	1.369	1.642	1.916	2.190	2.468	2.737	2.737	2.737
Net production (wine)	0.456	0.547	0.639	0.730	0.821	0.9125	0.912	0.912
Net production (beer)	1.369	1.642	1.916	2.190	2.464	2.7375	2.737	2.737
Add: Opening stock of wine	0	0.182	0.182	0.183	0.183	0.183	0.183	0.183
Add: Opening stock of beer	0	0.547	0.547	0.548	0.548	0.548	0.548	0.548
Less: Closing stock of wine	0.1825	0.182	0.182	0.183	0.183	0.183	0.183	0.183
Less: Closing stock of beer	0.5475	0.547	0.547	0.548	0.548	0.548	0.548	0.548
Sale quantity (wine)	0.004	0.008	0.009	0.010	0.011	0.013	0.013	0.013
Sale quantity (beer)	0.011	0.023	0.027	0.030	0.034	0.038	0.038	0.038
Dealer Price per litre of wine	2.3	2.3	2.3	2.3	2.3	2.3	2.3	2.3
Dealer Price per litre of beer	0.86	0.86	0.86	0.86	0.86	0.86	0.86	0.86
Sale value of wine	0.631	1.262	1.473	1.683	1.893	2.104	2.104	2.104
Sale value of beer	0.712	1.424	1.661	1.898	2.135	2.373	2.373	2.373

*In Table 4.21, the production quantity of purple sweet potato wine and purple sweet potato beer is calculated in million litres; dealer price per litre of purple sweet potato wine and beer is calculated in USD and sale value of purple sweet potato wine ard beer is calculated in million USD.

Table 8.9. Projected profitability statement (year end: 31st March).

	2020	2021	2022	2023	2024	2025	2026	2027
	Projected 50%	Projected 60%	Projected 70%	Projected 80%	Projected 90%	Projected 100%	Projected 100%	Projected 100%
Sales of purple sweet potato wine (million litres)	0.456	0.548	0.639	0.730	0.821	0.913	0.913	0.913
Sales of purple sweet potato beer (million litres)	1.369	1.643	1.916	2.190	2.464	2.738	2.738	2.738
Manufacturing cost of wine (million USD)	0.136	0.163	0.190	0.217	0.244	0.272	0.272	0.272
Manufacturing cost of beer (million USD)	0.425	0.510	0.595	0.680	0.765	0.850	0.850	0.850
Add: Opening stock of finished wine	–	0.183	0.183	0.183	0.183	0.183	0.183	0.183
Add: Opening stock of finished beer	–	0.548	0.548	0.548	0.548	0.548	0.548	0.548
Less: Closing stock of finished wine (million litres)	0.183	0.183	0.183	0.183	0.183	0.183	0.183	0.183
Less: Closing stock of finished beer (million litres)	0.548	0.548	0.548	0.548	0.548	0.548	0.548	0.548
Gross revenue from sale of wine (million USD)	1.052	1.262	1.473	1.683	1.893	2.104	2.104	1.052
Gross revenue From sale of beer (million USD)	1.186	1.424	1.661	1.898	2.135	2.373	2.373	1.186
Gross profit from sale of wine (million USD)	0.916	1.099	1.283	1.466	1.649	1.832	1.832	1.832
Gross profit from sale of beer (million USD)	0.761	0.913	1.066	1.218	1.370	1.522	1.522	1.522
Administrative and selling cost of wine (million USD)	0.434	0.414	0.394	0.373	0.353	0.343	0.343	0.343

Table 8.9 contd. ....

155

...*Table 8.9 contd.*

	2020	2021	2022	2023	2024	2025	2026	2027
	Projected 50%	Projected 60%	Projected 70%	Projected 80%	Projected 90%	Projected 100%	Projected 100%	Projected 100%
Administrative and selling cost of beer (million USD)	0.434	0.414	0.394	0.373	0.353	0.343	0.343	0.343
Net profit From wine (million USD)	0.482	0.686	0.889	1.092	1.296	1.489	1.489	1.489
Net profit from beer (million USD)	0.327	0.500	0.672	0.844	1.017	1.179	1.179	1.179
% of profit for Wine Sales	355	421	468	503	530	548	548	548
% of profit for beer sales	77	98	113	124	133	139	139	139
ROCE: (net profit before tax and interest/ capital Employed) for wine	172%	223%	275%	326%	377%	292%	292%	292%
ROCE: (net profit before tax and interest/ capital employed) for beer	128%	171%	214%	256%	299%	232%	232%	232%
Net profit before tax and interest for wine	0.612	0.795	0.978	1.161	1.345	1.528	1.528	1.528
Net profit before tax and interest for beer	0.457	0.609	0.761	0.913	1.066	1.218	1.218	1.218
Capital employed (Wine)= fixed asset-current liablity	0.356	0.356	0.356	0.356	0.356	0.524	0.524	0.524
Capital employed(beer)= fixed asset-current liablity	0.356	0.356	0.356	0.356	0.356	0.524	0.524	0.524

*In Table 4.22, sales and closing stock of purple sweet potato wine and purple sweet potato beer are in million litres while profits and selling cost are in million USD.

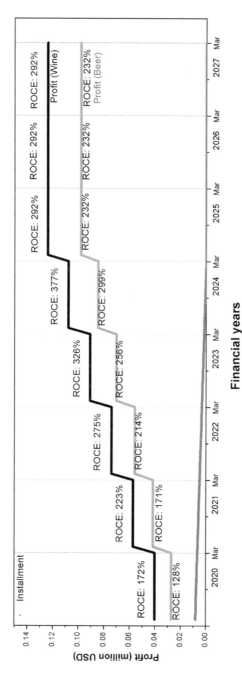

Financial years

Color version at the end of the book

Figure 8.3. Year wise projection of installment to be paid to bank and profit from sales of purple sweet potato wine and beer.

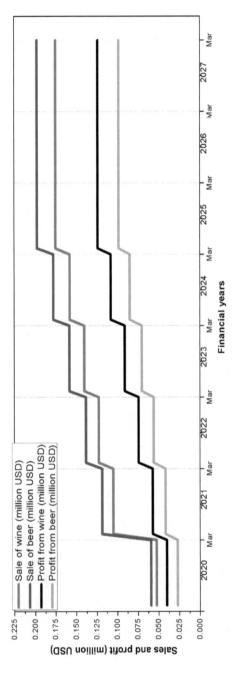

Figure 8.4. Year wise projection of sales versus profit for wine and beer.

Color version at the end of the book

Table 8.10. Term loan and installments.

Quantum: 0.83922 million USD								
Rate of interest: 12% per annum								
Year	Month	Opening balance	Availment	Repayment	Closing balance	Average balance	Interest	Annual interest
2020-21	Apr	0.839		0.014	0.825	0.832	0.0083	–
	May	0.825		0.014	0.811	0.818	0.0082	–
	Jun	0.811		0.014	0.797	0.804	0.0080	–
	Jul	0.797		0.014	0.783	0.790	0.0079	–
	Aug	0.783		0.014	0.769	0.776	0.0078	–
	Sep	0.769		0.014	0.755	0.762	0.0076	–
	Oct	0.755		0.014	0.741	0.748	0.0075	–
	Nov	0.741		0.014	0.727	0.734	0.0073	–
	Dec	0.727		0.014	0.713	0.720	0.0072	–
	Jan	0.713		0.014	0.699	0.706	0.0071	–
	Feb	0.699		0.014	0.685	0.692	0.0069	–
	Mar	0.685		0.014	0.671	0.678	0.0068	0.091
2021-22	Apr	0.671		0.014	0.657	0.664	0.0066	–
	May	0.657		0.014	0.643	0.650	0.0065	–
	Jun	0.643		0.014	0.629	0.636	0.0064	–
	Jul	0.629		0.014	0.615	0.622	0.0062	–

Table 8.10 contd. ....

...Table 8.10 contd.

Quantum: 0.83922 million USD

Rate of interest: 12% per annum

Year	Month	Opening balance	Availment	Repayment	Closing balance	Average balance	Interest	Annual interest
2021-22	Aug	0.615		0.014	0.601	0.608	0.0061	–
	Sep	0.601		0.014	0.587	0.594	0.0059	–
	Oct	0.587		0.014	0.573	0.580	0.0058	–
	Nov	0.573		0.014	0.559	0.566	0.0057	–
	Dec	0.559		0.014	0.545	0.552	0.0055	–
	Jan	0.545		0.014	0.532	0.539	0.0054	–
	Feb	0.532		0.014	0.518	0.525	0.0052	–
	Mar	0.518		0.014	0.504	0.511	0.0051	0.070
2022-23	Apr	0.504		0.014	0.490	0.497	0.0050	–
	May	0.490		0.014	0.476	0.483	0.0048	–
	Jun	0.476		0.014	0.462	0.469	0.0047	–
	Jul	0.462		0.014	0.448	0.455	0.0045	–
	Aug	0.448		0.014	0.434	0.441	0.0044	–
	Sep	0.434		0.014	0.420	0.427	0.0043	–
	Oct	0.420		0.014	0.406	0.413	0.0041	–
	Nov	0.406		0.014	0.392	0.399	0.0040	–

Year	Month							
2023-24	Dec	0.392		0.014		0.378	0.385	0.0038	–
	Jan	0.378		0.014		0.364	0.371	0.0037	–
	Feb	0.364		0.014		0.350	0.357	0.0036	–
	Mar	0.350		0.014		0.336	0.343	0.0034	0.050
	Apr	0.336		0.014		0.322	0.329	0.0033	–
	May	0.322		0.014		0.308	0.315	0.0031	–
	Jun	0.308		0.014		0.294	0.301	0.0030	–
	Jul	0.294		0.014		0.280	0.287	0.0029	–
	Aug	0.280		0.014		0.266	0.273	0.0027	–
	Sep	0.266		0.014		0.252	0.259	0.0026	–
	Oct	0.252		0.014		0.238	0.245	0.0024	–
	Nov	0.238		0.014		0.224	0.231	0.0023	–
	Dec	0.224		0.014		0.210	0.217	0.0022	–
	Jan	0.210		0.014		0.196	0.203	0.0020	–
	Feb	0.196		0.014		0.182	0.189	0.0019	–
	Mar	0.182		0.014		0.168	0.175	0.0017	0.030
2024-25	Apr	0.168		0.014		0.154	0.161	0.0016	–
	May	0.154		0.014		0.140	0.147	0.0015	–
	Jun	0.140		0.014		0.126	0.133	0.0013	–
	Jul	0.126		0.014		0.112	0.119	0.0012	–
	Aug	0.112		0.014		0.098	0.105	0.0010	–
	Sep	0.098		0.014		0.084	0.091	0.0009	–

Table 8.10 contd. ....

161

...Table 8.10 contd.

Quantum: 0.83922 million USD								
Rate of interest: 12% per annum								
Year	Month	Opening balance	Availment	Repayment	Closing balance	Average balance	Interest	Annual interest
2024-25	Oct	0.084		0.014	0.070	0.077	0.0008	–
	Nov	0.070		0.014	0.056	0.063	0.0006	–
	Dec	0.056		0.014	0.042	0.049	0.0005	–
	Jan	0.042		0.014	0.028	0.035	0.0003	–
	Feb	0.028		0.014	0.014	0.021	0.0002	–
	Mar	0.014		0.014	-	0.007	0.0001	0.010
Total				0.839			0.252	
Repayment + Interest							1.091	

Finally, the monthly premium would be paid by the company as per the estimation obtained from the data shown in Table 8.10. As evident from the table, by the end of the financial year 2025, the loan would be completely repaid, which indicates a viable business plan.

9

Summary and Conclusion

◇◇

Studies on the alcoholic fermentation of sweet potato have provided an optimized recipe and technology for the preparation of the three novel products, i.e., (1) anthocyanin rich wine (red wine), (2) herbal sweet potato wine and (3) anthocyanin rich beer having higher dietary importance.

Purple sweet potato is a special type of sweet potato containing high anthocyanin pigment in the root. The starch contents of purple sweet potato (root: water homogenized in 1:1 ratio) were enzymatically saccharified [using commercial thermostable enzymes Termamyl® (0.2%) and Dextrozyme® GA (1%)] to fermentable sugars and the filtrate was ameliorated with cane sugar to achieve 20°Brix, for subsequent fermentation into a red wine using 2% yeast (*Saccharomyces cerevisiae*) as starter culture. The wine had the following proximate compositions: TSS, 2.25°Brix; starch, 0.15 g/100 ml; total sugar, 1.35 g/100 ml; titratable acidity, 1.34 g tartaric acid/100 ml; phenol, 0.36 g (caffeic acid equivalent)/100 ml; anthocyanin, 55.09 mg/100 ml; tannin, 0.64 mg/100 ml; lactic acid, 1.14 mg/100 ml; ethanol, 9.33% (v/v) and pH of 3.61. DPPH scavenging activity of the wine was 58.95% at a dose of 250 μg/ml. Principal component analysis reduced the 11 original analytical and proximate variables (TSS, starch, total sugar, titratable acidity, phenol, anthocyanin, tannin, lactic acid, ethanol, pH, and DPPH) to four independent components, which accounted for 76.65% variations. Sensory evaluation was carried out by 16 trained panelists on various attributes such as clarity, colour, viscosity, odour, flavour, spritz and finish. Independent '*t*' test confirmed that all the sensory attributes of the purple sweet potato wine scored closely (within 10 % variation) to that of a commercial grape wine. The red wine produced contains essential antioxidants and acceptable sensory qualities.

A herbal purple sweet potato wine was prepared from the purple fleshed sweet potato and 18 medicinal plant parts (fruits of ink nut, Indian gooseberry, garlic cinnamon, leaves of holy basil, night jasmine, Malabar nut, roots of belladonna, asparagus, rhizome of ginger, etc.), by fermenting with wine yeast, *S. cerevisiae*. The starch present in the purple sweet potato was enzymatically saccharified [using commercial thermostable enzymes Termamyl® (0.2%) and Dextrozyme® GA (1%)] to fermentable sugars and the homogenized medicinal plant parts were mixed to it in desirable quantities, before subjected to fermentation. The herbal wine had the following compositions: TSS, 4.0°Brix; starch, 0.24 g/100 ml; total sugar, 0.95 g/100 ml; reducing sugar; 0.38 g/100 ml; titratable acidity, 1.25 g tartaric acid/100 ml; phenol, 0.19 g (caffeic acid equivalent)/100 ml; anthocyanin, 59.90 mg/100 ml; lactic acid, 1.92 mg/100 ml; ethanol, 8.61% (v/v) and pH of 3.34. DPPH scavenging activity of the wine was 51.35% at a dose of 250 µg/ml. The herbal wine thus prepared was presumed to contain the therapeutic and antioxidant properties of purple sweet potato as well as those of medicinal plant parts added as adjuncts. Principal component analysis reduced the 11 original analytical and proximate variables (TSS, reducing sugars, starch, total sugar, titratable acidity, pH, phenol, DPPH, lactic acid, ethanol, and anthocyanin) to four independent components, which accounted for 74.53% variations (PC1, 24.18%; PC2, 20.18%; PC3, 15.41%; PC4, 14.76%).

Beer was developed by the combination of purple sweet potato flakes and grist prepared from barley. Among the beer samples (prepared from 0, 30, 50, and 100% purple sweet potato flakes as adjuncts) the beer prepared with 30% purple sweet potato flakes as adjunct was the most preferred by the trained panelists. The beer is a light pink colour beverage because of its low anthocyanin content (2.31 mg/100 g). It is lighter in alcoholic content (3.8%), i.e., '*laser*' in type. The beer prepared with 30% purple sweet potato flakes had the following compositions: TSS, 3.57°Brix; starch, 0.77 g/100 ml; total sugar, 6.93 g/100 ml; specific gravity, 1.02; pH, 3.05; titratable acidity,

0.76 g tartaric acid/100 ml; lactic acid, 0.10 mg/100 ml; phenol, 0.21 mg/ml; DPPH scavenging activity, 3.15% at a dose of 250 µg/ml. PCA reduced the 11 biochemical parameters into 3 components (PC1–PC3). PC1, PC2 and PC3 accounted for 46.92, 27.52 and 21.25% of the total variations. Five analytical variables, i.e., TSS (–), pH (+), phenol (+), anthocyanin (+) and DPPH (+) were loaded heavily on PC1. The correlation between phenol (+), anthocyanin (+) and DPPH (+) indicated strong correlations among their attributes. Substantial factor loading of TA (+), LA (–) and ethanol (+) on PC2 and starch (+) and specific gravity (+) on PC3 was observed. The purple sweet potato beer has an advantage over the beer available in market (prepared from malt and adjuncts other than purple sweet potato) as it contains anthocyanin, a higher phenolic content and higher DPPH scavenging activity. The strong correlation among the anthocyanin pigment, phenol and the DPPH scavenging activity in the beer sample confirms the health attributes of the beer. The beer prepared from 30% purple sweet potato flakes + 70% grist was the best as recommended by the sensory panelists (mean value –7), followed by the beer prepared by 50% purple sweet potato flakes + 50% grist (mean value –6).

Technoeconomical and costing analysis study was conducted by assuming an integrated unit with production capacity of 2500 litres of wine per day and 7500 litres of beer per day. The detail report depicts that the project of an integrated winery-cum-brewery customized to use purple sweet potato as raw material is profitable as an innovative venture.

10

References

A profile of South African Sweet Potato Market Value Chain. 2017. Accessed from https://www.nda.agric.za/doaDev/sideMenu/Marketing/Annual%20Publications/Commodity%20Profiles/field%20crops/Sweet%20Potato%20Market%20Value%20Chain%20Profile%202017.pdf on 10.11.2012.

Adams, J. B. 2004. Raw materials quality and the texture of processed vegetables. *In*: D. Kilkast (ed.). Texture in Food, pp. 342–363. USA: Woodhead Publishing.

Adedotun, H., Adebowale, A. R., Olayiwola, I. O., Shittu, T. A. and Sanni, L. O. 2015. Production and quality evaluation of noodles from sweet potato starch. Journal of Culinary Science and Technology 13(1): 79–93.

Adeola, A. A. and Ohizua, E. R. 2018. Physical, chemical, and sensory properties of biscuits prepared from flour blends of unripe cooking banana, pigeon pea, and sweet potato. Food Science and Nutrition 6(3): 532–40.

Adesina, O. A., Oluwabunmi, K. E., Betiku, E., Fatuntele, L. T., Ayodele, O. A. and Adesanwo, C. A. 2014. Optimization of process variables for the production of oxalic acid from sweet potato starch hydrolyzate. Chemical and Process Engineering Research 18: 16–25.

Adeyosoye, O. I., Adesokan, I. A., Afolabi, K. D. and Ekeocha, A. H. 2010. Estimation of proximate composition and biogas production from *in vitro* gas fermentation of sweet potato (*Ipomea batatas*) and wild cocoyam (*Colocasia esculenta*) peels. African Journal of Environmental Science and Technology 4(6): 388–391.

Adthalungrong, A., Adthalungrong, C., Sirithanachareon, A. and Prachansit, A. 2014. Lactic Acid Production From Sweet Potato by *Lactobacillus Casei* Tistr 453. Bali, Indonesia: International Conference on Global Trends in Academic Research (June 2–3).

Agu, R. C., Bringhurst, T. A. Brosnan, J. M. and Jack, F. R. 2008. Effect of process conditions on alcohol yield of wheat, maize and other cereals. Journal of the Institute of Brewing 114(1): 39–44.

Ahmed, M., Akter, M. S. and Eun. J. B. 2010. Peeling, drying temperatures, and sulphite-treatment affect physicochemical properties and nutritional quality of sweet potato flour. Food Chemistry 121(1): 112–118.

167

Akoetey, W. 2015. Direct fermentation of sweet potato starch into lactic acid by Lactobacillus amylovorus: The prospect of an adaptation process. (Masters dissertation thesis in food science). Faytteville: University of Arkansas.

Akubor, P. I., Obio, S. O., Nwadomere, K. A. and Obiomah, E. 2003. Production and evaluation of banana wine. Plant Foods for Human Nutrition 58: 1–6.

Alves, E. J. 1999. A Cultura de Banana: Aspectos Tecnicos, Socio-economico e Agro-industrial., Brasília: DF, Embrapa – SPI.

Amagloh, F. K. and Coad, J. 2014. Orange-fleshed sweet potato-based infant food is a better source of dietary vitamin A than a maize—Legume blend as complementary food. Food and Nutrition Bulletin 35(1): 51–9.

Amerine, M. A. and Ough, C. 1980. Methods for Analysis of Musts and Wines. New York, NY: Wiley-Inter Science Publication.

Aroyeun, S. O., Olubamiwa, O. and Ogunjobi, M. A. K. 2005. Development of wine from infused tea leaves (*Cammellia sinensis*). British Food Journal 107(1): 34–41.

Arvanitoyannis, I. S. and Tzouros, N. E. 2005. Implementation of quality control methods in conjugation with chemometrics towards authentication of dairy products. Critical Reviews in Food Science and Nutrition 45: 231–249.

Austin, D. F. 1985. Sprits in the hills of Peru. Fairchild Tropical Garden Bulletins 40(2): 6–13.

Babu, A. S., Parimalavalli, R., Jagannadham, K. and Rao, J. S. 2015. Chemical and structural properties of sweet potato starch treated with organic and inorganic acid. Journal of Food Science and Technology 52(9): 5745–53.

Bajamo, M. F. and Young, T. W. 1992. Development of mashing profile for the use of microbial enzymes in brewing with raw sorghum (80%) and malted barley or sorghum malt (20%). Journal of Institute of Brewing 98: 515–523.

Bamforth, C. W. 2004. Beer Health and Nutrition. 9600 Garsington Road, Oxford OX4 2DQ: Blackwell Science Ltd.

Battcock, M. and Azam–Ali, S. 2001. Fermented Fruits and Vegetables: A Global Perspectives. Rome: Food and Agricultural Organization.

Bechoff, A., Dufour, D., Dhuique–Mayer, C., Marouze, C., Reynes, M. and Westby, A. 2009. Effect of hot air, solar and sun drying treatments on provitamin A retention in orange-fleshed sweet potato. Journal of Food Engineering 92(2): 164–71.

Berger, H. M., Scott, P. H., Kenward, C., Scott, P. I. and Kharton, B. A. 1979. Curd and whey protein in the nutrition of low birth weight babies. Archives of Disease in Childhood 54: 98–104.

Berkowitz, M. 1996. HYPERLINK: http://www.archaeology.org/9609/newsbriefs/wine.html, World's Earliest Wine. HYPERLINK: https://en.wikipedia.org/wiki/Archaeology_(magazine)Archaeology 49 (5).

Betiku, E. and Adesina, O. A. 2013. Optimization of sweet potato starch hydrolyzate production and its potential utilization as substrate for citric acid production. British Biotechnology Journal 3(2): 169–182.

Beuchat, L. R. 1994. Antimicrobial properties of spices and their essential oils. pp. 167–179. *In*: Y. M. Dillon and R. G. Board (eds.). Natural Antimicrobial Systems and Food Preservation. Oxon: CAB International.

Beuchat, L. R. and Golden, D. A. 1989. Antimicrobials occurring naturally in foods. Food Technology 43: 134–142.

Bhat, R. B. and Moskovitz, G. 2009. Herbal medicinal teas from South Africa. International Journal of Experimental Botany 78: 67–73.

Biswas, K., Chattopadhyay, I., Banerjee, R. K. and Bandyopadhyay, U. 2002. Biological activities and medicinal properties of neem (*Azadirachta indica*). Current Science 82(11): 1336–1345.

Bonciu, C. and Stoicescu, A. 2008. Obtaining and characterization of beers with cherries. Innovative Romanian Food Biotechnology 3: 23–27.

Bond, J. 2017. U.S. Sweet Potato Production Swells. Accessed from https://www.usda.gov/media/blog/2017/01/05/us-sweet-potato-production-swells on 10.10.2018.

Bridgers, E. N., Chinn, M. S. and Truong, V. D. 2010. Extraction of anthocyanins from industrial purple-fleshed sweet potatoes and enzymatic hydrolysis of residues for fermentable sugars. Industrial Crops and Products 32(3): 613–20.

Button K. 2015. Processing sweet potatoes into French fries. Thesis submitted in partial fulfillment of the degree of Master of Technology in Food Technology. Manhattan: Kansas State University.

Caetano, P. K., Mariano–Nasser, F. A., Mendonca,V. Z., Furlaneto, K. A., Daiuto, E. R. and Vieites, R. L. 2018. Physicochemical and sensory characteristics of sweet potato chips undergoing different cooking methods. Food Science and Technology 38(3): 434–40.

Capece, A., Romaniello, R., Siesto, G. and Romano, P. 2018. Conventional and non-conventional yeasts in beer production. Fermentation 4(2): 38.

Carvalho, G. B. M., Silva, D. P., Bento, C. V., Vicente, A. A., Teixeira, J. A., Felipa, M. G. A. and Silva, J. B. A. 2009. Banana as adjunct in beer production: Applicability and performance in fermentative parameters. Applied Biochemistry and Biotechnology 155: 356–365.

Cevallos–Casals, B. A. and Cisneros–Zevallos, L. A. 2003. Stoichiometric and kinetic studies of phenolic antioxidants from Andean purple corn and red-fleshed sweet potato. Journal of Agricultural and Food Chemistry 51: 3313–3319.

Chandrasekara, A. and Kumar, T. J. 2016. Roots and tuber crops as functional foods: a review on phytochemical constituents and their potential health benefits. International Journal of Food Science 2016. Article ID 3631647, http://dx.doi.org/10.1155/2016/3631647.

Choi, J. M., Kim, J. H. and Cho, E. J. 2010 b. Protective activity of purple sweet potato extract added soymilk fermented by *Bacillus subtilis* against oxidative stress. Food Science and Biotechnology 19: 457–462.

Choi, S. P., Nguyen, M. T. and Sim, S. J. 2010a. Enzymatic pretreatment of *Chlamydomonas reinhardtii* biomass for ethanol production. Bioresource Technology 104: 5330–5336.

Chowdhury, P. and Ray, R. C. 2007. Fermentation of jamun (*Syzgium cumini* L.) fruits to form red wine. ASEAN Food Journal 14: 15–23.

Ciani, M. and Maccarelli, F. 1998. Oenological properties of non-Saccharomyces yeasts associated with wine making. World Journal of Microbiology and Biotechnology 49: 283–288.

Claassen, M. R. and Lawless, H. T. 1992. A comparison of descriptive terminology systems for the sensory analysis of flavour defects in milk. Journal of Food Science 57: 596–600.

Collins, J. L., Ebah, C. B., Mount, J. R., Demott, B. J. and Draughon, F. A. 1991. Production and evaluation of milk-sweet potato mixtures fermented with yogurt bacteria. Journal of Food Science 56(3): 685–688.

Colthup, N. B., Daly, L. H. and Wiberly, S. E. 1990. Introduction to Infrared and Raman Spectroscopy (3rd edn). New York: Academic Press.

Contreras, A., Hidalgo, C., Henschke, P. A., Chambers, P. J., Curtin, C. and Varela, C. 2014. Evaluation of non-Saccharomyces yeasts for the reduction of alcohol content in wine. Applied and Environmental Microbiology 80(5): 1670–1678.

Coronel, P., Truong, V. D., Simunovic, J., Sandeep, K. P. and Cartwright, G. D. 2005. Aseptic processing of sweet potato purees using a continuous flow microwave system. Journal of Food Science 70(9): E531–6.

Cowan, M. M. 1999. Plant products as antimicrobial agents. Clinical Microbiology Reviews 12(4): 564–582.

Cutler, H. G. 1995. Natural product flavour compounds as potential antimicrobials, insecticides, and medicinals. Agro Food Industry Hi-Tech 6: 19–23.

Darias–Martin,J.,Lobo-Rodrigo,G.,Hernandez–Cordero,J.,Diaz–Diaz,E.andDiaz–Romero,C. 2003. Alcoholic beverages obtained from black mulberry. Food Technology and Biotechnology 41(2): 173–176.

Dash, P. K., Mohaptra, S., Swain, M. R. and Thatoi, H. 2017. Optimization of bioethanol production from saccharified sweet potato root flour by co-fermentation of *Saccharomyces cerevisiae* and *Pichia* sp. using OVAT and response surface methodologies. Acta Biologica Szegediensis 61(1): 13–23.

Debebe, A., Redi-Abshiro, M. and Chandravanshi, B. S. 2017. Non-destructive determination of ethanol levels in fermented alcoholic beverages using Fourier transform mid-infrared spectroscopy. Chemistry Central Journal 11(1). doi:10.1186/s13065-017-0257-5.

Donadini, G., Spigno, G., Fumi, M. D. and Pastori, R. 2008. Evaluation of ideal every Italian food and beer pairings with regular consumers and food and beverage experts. Journal of Institute of Brewing 114(4): 329–342.

Duarte, W. F., Dias, D. R., Pereira, G. V. M., Gervasio, I. M. and Schwan, R. F. 2009. Indigenous and inoculated yeast fermentation of gabiroba (*Campomanesia pubescens*) pulp for fruit wine production. Journal of Industrial Microbiology and Biotechnology 36(4): 557–569.

Eblinger, H. M. 2009. Handbook of Brewing: Process, Technology, Markets. Weinheim, Germany: Wiley-VCH Verlag.

El Sheikha, A. F. and Ray, R. C. 2017. Potential impacts of bioprocessing of sweet potato. Critical Reviews in Food Science and Nutrition 57(3): 455–71.

Engel, E. 1970. Exploration of the Chilca Canyon. Current Anthropology 11: 55–58.

Etim, M. U. and Etokakpan, O. U. 1992. Sorghum brewing using sweet potato enzymic flour to increase saccharification. World Journal of Microbiology and Biotechnology 8: 509–511.

European Beer Guide. 2006. Accessed from http://europeanbeerguide.net/frstats.htm on 05.06.2011.

FAO. 2017. Accessed from http://www.fao.org/faostat/en/#rankings/countries_by_commodity on 14.11.2018.

Fasina, O. O., Walter Jr., W. M., Fleming, H. P. and Simunovic, N. 2003. Viscoelastic properties of restructured sweet potato puree. International Journal of Food Science and Technology 38(4): 421–5.

Field, A. 2000. Discovering Statistics using SPSS for Windows: Advanced Techniques for the Beginner. London: Sage Publications.

Figueira, A. C., Makinde, O. and Vieira, M. C. 2011. Process optimisation of sweet potato (*Ipomoea batatas*) puree as an ingredient in a formulation of weaning food. Fruit, Vegetable and Cereal Science and Biotechnology. Global Science Books 2011 Mar: 25–34.

Ginting, E. 2013. Carotenoid extraction of orange-fleshed sweet potato and its application as natural food colorant. Jurnal Teknologi dan Industri Pangan 24(1): 81–88.

Ginting, E. and Yulifianti, R. 2015. Characteristics of noodle prepared from orange-fleshed sweet potato, and domestic wheat flour. Procedia Food Science 3: 289–302.

Gishen, M., Cozzolino, D. and Dambergs, R. G. 2006. The analysis of grapes, wine, and other alcoholic beverages by infrared spectroscopy. pp. 539–556. *In*: E. C. Y., Li-Chan, P. R. Griffiths and J. M. Chalmers (eds.). Handbook of Vibrational Spectroscopy. Chicester, UK: John Wiley and Sons Ltd.

Gonzalez, S. S., Barrio, E. and Querol, A. 2008. Molecular characterization of new natural hybrids of *Saccharomyces cerevisiae* and *S. kudriavzevii* in brewing. Applied and Environmental Microbiology 74: 2314–2320.

Goode, D. L., Wijngaard, H. H. and Arendt, E. L. 2005. Mashing with unmalted barley-impact of malted barley and commercial enzyme (*Bacillus* spp.) additions. Master Brewers Association of the Americas 42(3): 184–198.

Gore, J. 2008. Comprehensive Study of the Indian Wine Market. Washington D.C: JBC International Inc.

Gorinstein, S., Caspi, A., Zemser, M. and Trakhtenberg, S. 2000. Comparative contents of some phenolics in beer, red and white wines. Nutrition Research 20: 131–139.

Govindasamy, S., Campanella, O. H. and Oates, C. G. 1997. Enzymatic hydrolysis and saccharification optimization of sago starch in a twin screw extruder. Journal of Food Engineering 32: 427–446.

Hair, J. F., Anderson, R. E., Tatham, R. L. and Black, W. C. 1998. Multivariate Data Analysis (5th edn). Upper Saddle River, NJ: Pearson Education Inc.

Han, K. H., Shimada, K., Sekikawa, M. and Fukushima, M. 2007. Anthocyanin-rich red potato flakes affect serum lipid peroxidation and hepatic SOD mRNA level in rats. Bioscience Biotechnology and Biochemistry 71: 1356–1359.

Harborne, J. B. 1998. Phytochemical Methods: A Guide to Modern Techniques of Plant Analysis (3rd edn). London: Chapman and Hall.

Hatano, T., Kagawa, H., Yasuhara, T. and Okuda, T. 1988. Two new flavonoids and other constituents in licore root: their relative astringency and radical scavenging affects. Chemical and Pharmaceutical Bulletin 36: 1090–2097.

Hayashi, K., Ohara, N. and Tsukui, A. 1996. Stability of anthocyanins in various vegetables and fruits. Food Science and Technology International 2: 30–33.

Hayashi, K., Mori, M., Knox, Y. M., Suzutan, T., Ogasawara, M., Yoshida, I., Hosokawa, L., Tsukui, A. and Azuma, M. 2003. Anti- influenza virus activity of a purple-fleshed potato anthocyanin. Food Science and Technology Research 9: 242–244.

Hoover, M. W., Walter Jr., W. M. and Giesbrecht, F. G. 1983. Method of preparation and sensory evaluation of sweet potato pattis. Journal of Food Science 48(5): 1568–1569.

Horton, D., Prain, G. and. P. Gregory. 1989. High level investment returns for global sweet potato research and development. *In*: F. Ofori and S. K Hahn (eds.). Tropical Root Crops in Developing Economy. Accra, Ghana: C.I.P. Circular 17.

Horvathova, K., Vachalkova, A. and Novotny, L. 2001. Flavonoids as chemoprotective agents in civilization diseases. Neoplasma 48: 435–441.

http://in.one.un.org/un-priority-areas-in-india/nutrition-and-food-security/. Accessed on 14.03.2019.

http://www.fao.org/3/am866e/am866e00.pdf). Accessed on 14.03.2019.

http://www.idco.in/2009/landrate.aspx. Accessed on 3.12.2018.

https://www.chemours.com/Teflon/en_US/assets/downloads/pdf/201407_Sweet_Potato_Waffle.pdf. Accessed on 14.03.2019.

https://www.kirinholdings.co.jp/english/news/2018/0809_01.html Accessed on 3.11.2018.

https://www.nri.org/development-programmes/root-and-tuber-crops-in-development/overview Accessed on 4.10.2018.

https://www.statista.com/statistics/240638/wine-production-in-selected-countries-and-regions/ Accessed on 12.10.2018.

https://www.statista.com/statistics/812343/global-sweet-potato-production/Assessed on 14.11.2018.

https://www.worldatlas.com/articles/world-leaders-in-sweet-potato-production.htm Accessed on 10.11.2018.

171

Huaman, Z. 1992. Systematic botany and morphology of the sweet potato plant. pp. 22. Technical Information Bulletin (No-25). Lima, Peru: International Potato Center.

Huang, C. L., Liao, W. C., Chan, C. F. and Lai, Y. C. 2010. Optimization for the anthocyanin extraction from purple sweet potato roots, using response surface methodology. Journal of Taiwan Agricultural Research 59(3): 143–50.

Husain, A., Virmani, O. P., Popli, S. P., Misra, L. N., Gupta, M. M., Srivastava, G. N., Abraham, Z. and Singh, A. K. 1992. Dictionary of Indian Medicinal Plants. Lucknow, India: Central Institute of Medicinal and Aromatic Plants.

Ifeoluwa, A. K. 2017. Optimizing alpha amylase production from locally isolated *Aspergillus* species using selected agro wastes as substrates (Masters dissertation in microbiology). Nigeria: Babcock university.

Ishiwu, C. N., Nkwo, V. O., Iwouno, J. O, Obiegbuna, J. E. and Uchegbu, N. N. 2014. Optimization of taste and texture of biscuit produced from blend of plantain, sweet potato and malted sorghum flour. African Journal of Food Science 8(5): 233–238.

Jackson, R. S. 2017. Innovations in winemaking. pp. 617–662. In: M. R. Kosseva, V. K. Joshi and P. S. Panesar (eds.). Science and Technology of Fruit Wine Production. San Diego: Elsevier.

Jain, S. K. 1983. Medicinal Plants. New Delhi, India: National Book Trust.

Jemziya, M. B. and Mahendran, T. 2017. Physical quality characters of cookies produced from composite blends of wheat and sweet potato flour. Ruhuna Journal of Science 8: 12–23.

Joe, A., Vinson, J. J., Jihong, Y., Yousef, D., Xiquan, L., Mamdouh, S., John, P. and Sai, S. 1999. Vitamins and especially avonoids in common beverages are powerful *in vitro* antioxidants which enrich lower density lipoproteins and increase their oxidative resistance after *ex vivo* spiking in human plasma. Journal of Agricultural and Food Chemistry 47: 2502–2504.

Joseah, N. 2011. Production of a precooked complementary baby food based on orange fleshed sweet potato, finger millet, soybean and peanut flours. BSC Dissertation thesis, Kenya: University of Nairobi.

Joy, P. P., Thomas, J., Mathew, S. and Skaria, B. P. 1998. Medicinal Plants. Ernakulam, India: Kerala Agricultural University

Julianti, E., Rusmarilin, H. and Yusraini, E. 2017. Functional and rheological properties of composite flour from sweet potato, maize, soybean and xanthan gum. Journal of the Saudi Society of Agricultural Sciences 16(2): 171–177.

Jumirah, L. Z. and Lubis, Z. 2018. The composition of nutritious biscuits of sweet potato and tempe flour enriched with vitamin a of red palm oil. IIOAB Journal 9(1): 1–6.

Kale, R. V. 2017. Isolation, modification and characterization of starch from sweet potato (*Ipomoea batatas* L.) and its exploration in gulabjamun and ice-cream (Doctoral dissertation). Parbhani, India: Vasantrao Naik Marathwada Krishi Vidyapeeth.

Kandasamy, S., Kavitake, D. and Shetty, P. H. 2018. Lactic acid bacteria and yeasts as starter cultures for fermented foods and their role in commercialization of fermented Foods. pp. 25–52. In: S. K. Panda and P. K. Shetty (eds.). Innovations in Technologies for Fermented Food and Beverage Industries. Switzerland: Springer.

Karovicova, J., Kohajdovaz, Z. and Hybenova, E. 2001. Using of multivariate analysis for evaluation of lactic acid fermented cabbage juices. Chemical Papers 56(4): 267–274.

Keys, D. 2003. HYPERLINK: https://www.independent.co.uk/news/science/now-thats-what-you-call-a-real-vintage-professor-unearths-8000yearold-wine-84179.html. Now that's what you call a real vintage: professor unearths 8,000-year-old wine. HYPERLINK:https://en.wikipedia.org/wiki/The_Independent"The Independent (published in 28th December 2003).

Khoo, H. E., Azlan, A., Tang, S. T. and Lim, S. M. 2017. Anthocyanidins and anthocyanins: Colored pigments as food, pharmaceutical ingredients, and the potential health benefits. Food and Nutrition Research 61(1): 1361779. doi:10.1080/16546628.2017.1 361779.

Kilcast, D. and Subramanian, P. 2000. The Stability and Shelf-life of Food. Cambridge: Woodhead Publishing Limited.

Kim, S. J., Rhim, J. W., Lee, L. S. and Lee, J. S. 1996. Extraction and characteristics of purple sweet potato pigment. Korean Journal of Food Science and Technology 28(2): 345–51.

Kobayashi, T., Tang, Y., Urakami, T., Morimura, S. and Kida, K. 2014. Digestion performance and microbial community in full-scale methane fermentation of stillage from sweet potato-shochu production. Journal of Environmental Sciences 26(2): 423–31.

Kohajdova, Z., Karovicova, J. and Greifova, M. 2007. Analytical and organoleptic profiles of lactic acid-fermented cucumber juice with addition of onion juice. Journal of Food and Nutrition Research 46(3): 105–111

Kumar, A., Duhan, J. S., Gahlawat, S. and Gahlawat, S. K. 2014. Production of ethanol from tuberous plant (sweet potato) using *Saccharomyces cerevisiae* MTCC-170. African Journal of Biotechnology 13(28): 2874–2883.

Kumar, K. K., Swain, M. R., Panda, S. H., Sahoo, U. C. and Ray, R. C. 2008. Fermentation of litchi (*Litchi chinensis* Sonn.) fruits into wine. Food 2: 43–47.

Lachenmeier, D. W. 2007. Rapid quality control of spirit drinks and beer using multivariate data analysis of Fourier transform infrared spectra. Food Chemistry 101(2): 825–832.

Langstaff, S. A., Guinard, J. X. and Lewis, M. J. 1991. Instrumental evaluations of the mouth feel of beer and correlation with sensory evaluation. Journal of Institute of Brewing 97: 427–433.

Lareo, C., Ferrari, M. D., Guigou, M., Fajardo, L., Larnaudie, V., Ramirez, M. B. and Martínez-Garreiro, J. 2013. Evaluation of sweet potato for fuel bioethanol production: Hydrolysis and fermentation. Springerplus 2(493): 1–11.

Laryea, D., Wireko-Manu, F. D. and Oduro, I. 2018. Formulation and characterization of sweetpotato-based complementary food. Cogent Food and Agriculture 4(1): 1–5.

Lebot, V. 2008. Tropical Root and Tuber Crops, UK: CABI Publishing.

Lee, Y. and Kennedy, L. 2016. Determining the effect of NAFTA on the North American sweet potato market. In: 2016 Annual Meeting, July 31–August 2, 2016, Boston, Massachusetts (No. 235429): Agricultural and Applied Economics Association.

Li, J., Zhang, L. and Liu, Y. 2013. Optimization of extraction of natural pigment from purple sweet potato by response surface methodology and its stability. Journal of Chemistry 2013: Article ID 590512, 1–5.

Li, L., Liao, C. W. and Chin, L. 1994. Variability in test and physico-chemical properties and its breeding implications in sweet potato (*Ipomoea batatas*). Journal of the Agricultural Association of China New Series 165: 19–23.

Lilly, M., Lambrechts, M. G. and Pretorius, I. S. 2000. Effect of increased yeast alcoholacetyl transferase activity of flavour profiles of wine and distillates. Applied Environmental Microbiology 66: 744–753.

Lim, S., Xu, J., Kim, J., Chen, T. Y., Su, X., Standard, J. and Tomich, J. 2013. Role of anthocyanin-enriched purple-fleshed sweet potato p40 in colorectal cancer prevention. Molecular Nutrition and Food Research 57(11): 1908–1917.

Lu, L. Z., Zhou, Y. Z., Zang, Y. Q., Ma, Y. L., Zhou, L. X., Li, L., Zhou, Z. Z. and He, T. Z. 2010. Anthocyanin extracts from purple sweet potato by means of microwave baking and acidified electrolysed water and their antioxidation *in vitro*. International Journal of Food Science and Technology 45: 1378–1385.

Maetzu, L., Andueza, S., Ibanez, C., Pena, M. P., Bello, J. and Cid, C. 2001. Multivariate methods for characterization and classification of espresso coffees from different

Botanical varieties and types of roast by foam, taste, and mouthfeel. Journal of Agriculture and Food Chemistry 49: 4743–4747.

Mahadevan, A. and Sridhar, R. 1998. Methods in Physiological Plant Pathology (5th edn). Madras: Sivakami Publication.

Mais, A. 2008. Utilization of sweet potato starch, flour and fibre in bread and biscuits: Physico-chemical and nutritional characteristics. Thesis submitted in partial fulfillment of the the the degree of Master of Technology in Food Technology. New Zealand: Massey University.

Matthias, O. C. 2013. Optimization of α-amylase and glucoamylase production from three fungal strains isolated from Abakaliki, Ebonyi State. European Journal of Experimental Biology 3(4): 26–34.

Mesta, S. 2005. Beer Production from Sweet Sorghum Grains. Dharwad: University of Agricultural Sciences.

Minh, N. P. 2015. Factors Affecting to β-Caroten Extraction from Sweet Potato International Journal of Pure and Applied Bioscience 3(2): 396–399.

Mochizuki, M., Yamazaki, S., Kano, K. and Ikeda, T. 2002. Kinetic analysis and mechanistic aspects of autoxidation of catechins. Biochimica et Biophysica Acta 1569: 35–44.

Mohanty, S., Ray, P., Swain, M. R. and Ray, R. C. 2006. Fermentation of cashew (*Anacardium occidentale* L.) apple to wine. Journal of Food Processing and Preservation 30: 314–322.

Mohapatra, S., Panda, S. H., Sahoo, S. K., Sivakumar, P. S. and Ray, R. C. 2007. β-Carotene rich sweet potato curd: production, nutritional and proximate composition. International Journal of Food Science and Technology 42: 1305–1314.

Moorthy, S. N. 2004. Tropical sources of starches. pp. 321–359. *In*: (ed.). A. Eliasson. Starch in Food. England: Woodhead Publishing Limited.

Moorthy, S. N. and Shanavas, S. 2010. Sweet potato starch. *In*: R. C. Ray and K. I. Tomlins (eds.). Sweet Potato: Post Harvest Aspects in Food, New York: Nova Science Publishers, Inc.

Moorthy, S. N. 2010. Physicochemical characterization of selected sweet potato cultivars and their starches. International Journal of Food Properties 13: 1280–1289.

Mu, T. H., Li, P. G. and Sun, H. N. 2017. Bakery products and snacks based on sweet potato. *In*: H. K. Sharma, N. Y. Njintang, R. S. Singhal and P. Kaushal (eds.). Tropical Roots and Tubers: Production, Processing and Technology, Ist edition, West Sussex, UK: John Wiley and Sons Ltd.

Mulderji, R. 2016. http://www.freshplaza.com/article/2163603/overview-global-sweet-potato-market/ Accessed on 10.08.2018.

Mulderji, R. 2017. http://www.freshplaza.com/article/185147/overview-global-sweet-potato-market/ Accessed on 10.08.2018.

Muniz, C. R., Borges, M. F. and Freire, F. C. O. 2008. Tropical and subtropical fruit fermented beverages. pp. 35–70. *In*: R. C. Ray and O. P. Ward (eds.). Microbial Biotechnology in Horticulture, Vol. 3, New Hampshire: Science Publishers.

Murray, J. M. and Delahunty, C. M. 2000. Mapping consumer preference for the sensory and packaging attributes of cheddar cheese. Food Quality and Preference 11: 419–435.

Nagarajan, R., Mehrotra, R. and Bajaj, M. M. 2006. Qualitative analysis of methanol, an adulterant in alcoholic beverages, using attenuated total reflectance spectroscopy. Journal of Scientific and Industrial Research 65: 416–419.

Nandutu, A. M. and Howell, N. K. 2009. Nutritional and rheological properties of sweet potato based infant food and its preservation using antioxidants. African Journal of Food Agriculture, Nutrition and Development 9(4): 1076–1090.

Nedunchezhiyan, M. and Byju, G. 2005. Effect of planting season on growth and yield of sweet potato (*Ipomoea batatas* L.) varieties. Journal of Root Crops (31)2: 111–114.

Nedunchezhiyan, M. and Ray, R. C. 2010. Sweet potato growth, development, production and utilization. Pp.2–26. *In*: R. C. Ray and K. I. Tomlins (eds.). Sweet Potato: Post Harvest Aspects in Food Feed and Industry. New York, USA: Nova Publishers.

Nelson, S. C. and Elevitch, C. 2011. Farm and Forestry Production and Marketing Profile for Sweet Potato (*Ipomoea batatas* L.). Hawaii: University of Hawaii Press.

Nicanuru, C., Laswai, H. S. and Sila, D. N. 2015. Effect of sun-drying on nutrient content of orange fleshed sweet potato tubers in Tanzania. Sky Journal of Food Science 4(7): 91–101.

Nitar, N. and Stevens, W. F. 2002. Production of fungal chitosan by solid state fermentation followed by enzymatic extraction. Biotechnology Letters 24: 131–134.

Nogueira, A. C., Sehn, G. A., Rebellato, A. P., Coutinho, J. P., Godoy, H. T., Chang, Y. K., Steel, C. J. and Clerici, M. T. 2018. Yellow sweet potato flour: Use in sweet bread processing to increase β-carotene content and improve quality. Annals of the Brazilian Academy of Sciences 90(1): 283–93.

Norman, M. J. T., Pearson, C. J. and Searle, P. G. E. 1984. The Ecology of Tropical Food Crops. Cambridge: Cambridge University Press.

Odebode, S. O., Egeonu, N. and Akoroda, M. O. 2008. Promotion of sweet potato for the food industry in Nigeria. Bulgarian Journal of Agricultural Science 14(3): 300–308.

Odenigbo, A., Rahimi, J., Ngadi, M., Wees, D., Mustafa, A. and Seguin, P. 2012. Quality changes in different cultivars of sweet potato during deep-fat frying. Journal of Food Processing and Technology 3(5): 3–7.

Ogbonna, C. N., Nnaji, O. B. and Chioke, O. J. 2018. Isolation of amylase and cellulase producing fungi from decaying tubers and optimization of their enzyme production in solid and submerged cultures. International Journal of Biotechnology and Food Science 6(1): 9–17.

Okafor, N. 2006. Microbial bioconversions of agri-horticultural produces into alcoholic beverages—global scenario. pp. 411–452. *In*: R. C. Ray (ed.). Microbial Biotechnology in Agriculture and Aquaculture (Vol II). Enfield, NH, USA: Science Publishers.

Okigibo, R. N. and Obire, O. 2009. Mycoflora and production of wine from fruits of soursop (*Annona muricata* L.). International Journal of Wine Research 1: 1–9.

Olawoye, B., Adeniyi, D. M., Oyekunle, A. O., Kadiri, O. and Fawale, S. O. 2017. Economic evaluation of cookie made from blend of Brewers' spent grain (BSG), groundnut cake and sorghum flour. Open Agriculture 2(1): 401–410.

Oluwalana, I. B., Malomo, S. A. and Ogbodogbo, E. O. 2012. Quality assessment of flour and bread from sweet potato wheat composite flour blends. International Journal of Biological and Chemical Sciences 6(1): 65–76.

O'Manohy, M. 1986. Sensory Evaluation of Food. New York, NY: Marcel Dekker, Inc.

Onabanjo, O. O. and Ighere, D. A. 2014. Nutritional, functional and sensory properties of biscuit produced from wheat-sweet potato composite. Journal of Food Technology Research 1(2): 111–121.

Owori, C. and Hagenimana, V. 1998. Development of sweet potato snack products in rural areas: case study of Lira district, Uganda. pp. 11–17. *In*: Root Crops in the 21st Century: Proceedings. 7. Benin: Triennial Symposium of the International Society for Tropical Root Crops-Africa Branch (ISTRCAB).

Owuama, C. L. 1997. Sorghum: A cereal with lager brewing potential. World Journal of Microbiology and Biotechnology 13: 253–260.

Pagana, I. 2012. Lactic acid production using sweet potato processing waste (Masters dissertation thesis in food science). Faytteville: University of Arkansas.

Palmer, J. J. 2006. How to Brew. NE, Maston: Brewers Publication.

175

Panda, S. H., Naskar, S. K. and Ray, R. C. 2006. Production, proximate and nutritional evaluation of sweet potato curd. Journal of Food Agriculture and Environment 4: 124–127.

Panda, S. H. and Ray, R. C. 2007. Lactic Acid Fermentation of β-carotene rich sweet potato (*Ipomoea batatas* L.) into lacto-juice. Plant Food for Human Nutrition 62: 65–70.

Panda, S. H., Parmanick, M. and Panda, S. H. 2007. Lactic acid fermentation of sweet potato (*Ipomoea batatas* L.) into pickles. Journal of Food Processing and Preservation 31: 83–10.

Panda, S. H., Panda, S., Sivakumar, P. S. and Ray, R. C. 2009a. Anthocyanin rich sweet potato lacto-pickle: production, nutritional and proximate composition. International Journal of Food Science and Technology 44: 445–455.

Panda, S. H., Naskar, S. K., Sivakumar, P. S. and Ray, R. C. 2009b. Lactic acid fermentation of anthocyanin-rich sweet potato (*Ipomoea batatas* L.) into lacto-juice. International Journal of Food Science and Technology 44: 28–296.

Panda, S. K., Sahu, U. C., Behera, S. K. and Ray, R. C. 2014a. Fermentation of sapota (*Achras sapota* Linn.) fruits to functional wine. Nutrafoods 13(4): 179–186.

Panda, S. K., Sahu, U. C., Behera, S. K. and Ray, R. C. 2014b. Bio-processing of bael (*Aegle marmelos* L.) fruits into wine with antioxidants. Food Bioscience 5: 34–41.

Panda, S. K. and Ray, R. C. 2016. Fermented foods and beverages from roots and tubers. pp. 225–252. In: H. K. Sharma, N. Y. Njintang, R. S. Singhal and P. Kaushal (eds.). Tropical Roots and Tubers: Production, Processing and Technology, Ist edition, West Sussex, UK: John Wiley and Sons Ltd.

Panda, S. K., Behera, S. K., Sahoo, U. C., Ray, R. C., Kayitesi, E. and Mulaba-Bafubiandi, A.F. 2016a. Bioprocessing of jackfruit (*Artocarpus heterophyllus* L.) pulp into wine: technology, proximate composition and sensory evaluation. African Journal of Science Technology Innovation and Development 8(1): 27–32.

Panda, S. K., Mishra, S. S., Kayitesi, E. and Ray, R. C. 2016b. Microbial-processing of fruit and vegetable wastes for production of vital enzymes and organic acids: Biotechnology and scopes. Environmental Research 146: 161–172.

Panda, S. K., Ray, R. C, Mishra, S. S. and Kayitesi, E. 2018. Microbial processing of fruit and vegetable wastes into potential biocommodities: A review. Critical Reviews in Biotechnology 38(1): 1–6.

Paolucci, J. D., Belleville, M. P., Zakhia, N. and Rios, G. M. 2000. Kinetics of cassava starch hydrolysis with Termamyl enzyme. Biotechnology and Bioengineering 6: 71–77.

Peng, C. A., Yu, F. and Ma, Y. 2003. Preparation of health sweet potato bread. Journal of Wuhu Professional Technology College 4: 74–76.

Pereira, C. R., Resende, J. T., Guerra, E. P., Lima, V. A., Martins, M. D. and Knob, A. 2017. Enzymatic conversion of sweet potato granular starch into fermentable sugars: Feasibility of sweet potato peel as alternative substrate for α-amylase production. Biocatalysis and Agricultural Biotechnology 11: 231–8.

Perez, I. C., Mu, T. H., Zhang, M. and Ji, L. L. 2017. Effect of heat treatment to sweet potato flour on dough properties and characteristics of sweet potato-wheat bread. Food Science and Technology International 23(8): 708–15.

Perez, R. H. and Tan, J. D. 2006. Production of acidophilus milk enriched with purees from coloured sweet potato (*Ipomoea batatas* L.) varieties. Annals of Tropical Research 28: 70–85.

Peters, L. 2018. https://www.bustle.com/p/a-beer-wine-hybrid-exists-now-so-all-the-indecisive-humans-of-the-world-can-finally-find-peace-8442084. Accessed on 3.12.2018.

176

References

Phillips, L. G., McGiff, M. L., Barbano, D. M. and Lawless, H. T. 1995. The influence of fat on the sensory properties, viscosity, and color of lowfat milk. Journal of Dairy Science 78: 1258–1266.

Piaxao, N., Perestrelo, R., Marques, J. C. and Camara, J. S. 2007. Relationship between antioxidant capacity and total phenolic content of red, rose and white wines. Food Chemistry 105: 204–214.

Pongparnchedtha, T. and Suwanvisolkij, S. 2011. The Production of *Aloe vera* wine. http:// aseanfood.info/scripts/count_article.asp. Assessed on 9 June 2011.

Prakash, P. and Gupta, N. 2005. Therapeutic uses of *Ocimum sanctum* Linn (tulsi) with a note on eugenol and its pharmacological actions: a short review. Indian Journal of Physiology and Pharmacology 49(2): 125–131.

Ramasami, P., Jhaumeer–Laulloo, S., Cadet, F., Rondeau, P. and Soophul, Y. 2005. Quantification of alcohol in beverages by density and infrared spectroscopy methods. International Journal of Food Sciences and Nutrition 56(3): 177–83.

Ray, R. C. and Balagopalan, C. 1997. Post-harvest spoilage of sweet potato. pp. 31. India: Technical Bulletin (CTCRI) No. 17.

Ray, R. C. and Ravi, V. 2005. Postharvest spoilage of sweet potato in tropics and control measures. Critical Reviews in Food Science and Nutrition 45: 623–644.

Ray, R. C. and Ward, O. P. 2006. Microbial Biotechnology in Agriculture and Aquaculture, Vol. 2. New Hampshire: Science Publishers.

Ray, R. C. and Panda, S. H. 2007. Lactic acid fermentation of fruits and vegetables: An overview. pp. 155–188. *In*: Palino, M. V. (ed.). Food Microbiology Research Trends. New York: Nova Science Publishers.

Ray, R. C. and Naskar, S. K. 2008. Bioethanol production from sweet potato (*Ipomoea batatas* L.) by enzymatic liquefaction and simultaneous saccharification and fermentation (SSF) process. Dynamic Biochemistry, Process Biotechnology and Molecular Biology 2(1): 47–49.

Ray, R. C. and Tomlins, K. I. 2010. Sweet Potato: Post-harvest Aspects in Food, Feed and Industry. New York: Nova Science Publishers, Inc.

Reddy, L. V. A. and Reddy, O. V. S. 2005. Production and characterization of wine from mango fruit (*Mangifera indica* L.). World Journal of Microbiology Biotechnology 21: 1345–1350.

Rice-Evans, C. A. and Packer, L. 1998. Flavonoids in Health and Disease. New York: Marcel Dekker.

Rincon-Leon, F. 2003. Functional foods. pp. 2827–32. *In*: L. C. Trugo and P. M. Finglas (eds.). Encyclopedia of Food Sciences and Nutrition (Vol. 5) (2nd edn). London: Academic Press.

Rodriguez-Saona, L. E. and Wrolstad, R. E. 2001. Extraction, isolation, and purification of anthocyanins. *In*: Wrolstad, R. E. (ed.). Current Protocols in Food Analytical Chemistry. Indianapolis: John Wiley & Sons Inc.

Rojas, V., Gil, J. V., Pinaga, F. and Manzanares, P. 2001. Acetate ester formation in wine by mixed cultures in laboratory fermentations. International Journal of Food Microbiology (86)1: 181–188.

Romano, P., Suzzi, G., Comi, G. and Zironi, R. 1992. Higher alcohol and acetic acid production by apiculate wine yeasts. Journal of Applied Bacteriology 73: 126–130.

Rudgley, R. 1993. The Alchemy of Culture: Intoxicants in Society. London: British Museum Press.

Russell, I. and Kellershohn, J. 2018. Advances in technology and new product development in the beer, wine, and spirit industry. pp. 89–104. *In*: S. K. Panda and P. K. Shetty (eds.). Innovations in Technologies for Fermented Food and Beverage Industries. Switzerland: Springer.

Sahu, U. C., Panda, S. K., Mohapatra, U. B. and Ray, R. C. 2013. Preparation and evaluation of wine from tendu (*Diospyros melanoxylon* L.). International Journal of Food and Fermentation Technology 2(2): 167–178.

Saigusa, N., Terahara, N. and Ohba, R. 2005. Evaluation of DPPH-radical scavenging activity and antimutagenicity and analysis of anthocyanins in an alcoholic fermented beverage produced from cooked or raw purple-fleshed sweet potato (*Ipomoea batatas* cv. Ayamurasaki) roots. Food Science Technology Research 11: 390–394.

Sanchez–Moreno, C., Larrauri, J. A. and Saura-Calixto, F. 1998. A procedure to measure the antiradical efficiency of polyphenols. Journal of the Science of Food and Agriculture 76: 270–276.

Sardar, A. A., Ali, S., Anwar, S. and Haq, I. U. 2007. Citric acid fermentation of hydrolysed sweet potato starch by *Aspergillus niger*. Pakistan Journal of Biotechnology 4(1-2): 119–25.

Sarkar, S., Kulia, R. K. and Misra, A. K. 1996. Organoleptic, microbiological and chemical quality of misti dahi sold in different districts of West Bengal, India. Journal of Dairy Science 49: 54–61.

Sichel, G., Corsaro, C., Scalia, M., diBilio, A. J. and Bonomo, R. P. 1991. *In vitro* scavenger activity of some flavonoids and melanins against O_2. Free Radical Biology and Medicine 11: 1–8.

Sim, C. O., Hamdan, M. R., Ismail, Z. and Ahmad, M. N. 2004. Assessment of herbal medicines by chemometrics-assisted interpretation of FT-IR spectra. Journal of Analytica Chimica Acta, Assessed from http://www.camo.com/downloads/resourses/applicatons_notes/Assessment of herbal medicines by chemometrics-assisted interpretation of FT-IR spectra on 5th May 2011.

Singh, A. K., Lawrence, R., Jeyakumar, E. G. and Ramteke, P. W. 2014. Development of a solid-state fermentation process for production of bacterial α-Amylase from agro-byproducts and its optimization. International Journal of Bioinformatics and Biological Sciences 2(3): 201–226.

Singh, R. S. and Kaur, P. 2009. Evaluation of litchi juice concentrate for the production of wine. Natural Product Radiance 8(4): 386–391.

Singh, S., Raina, C. S., Bawa, A. S. and Saxena, D. C. 2004. Sweet potato-based pasta product: optimization of ingredient levels using response surface methodology. International Journal of Food Science and Technology 39(2): 191–200.

Soleas, G. L., Diamandis, E. P. and Goldberg, D. M. 1997. Wine as a biological fluid: History, production, and role in disease prevention. Journal of Clinical Laboratory Analysis 11: 287–313.

Soni, S. K., Bansal, N. and Soni, R. 2009. Standardization of conditions for fermentation and maturation of wine from Amla (*Emblica officinalis* Gaertn.). Natural Product Radiance 8(4): 436–444.

Sripanomtanakorm, S. and Siriboo, N. 1999. Study on the use of sweet potato flour in cake. AU Journal of Technology 3(1): 42–46.

Stevens, J. 1992. Applied multivariate statistics for the social sciences. Hillsdale: Lawrence Erlbaum Associates Inc.

Stuart, B. H. 2004. Infrared Spectroscopy: Fundamentals and Applications. Chichester, UK: John Wiley and Sons Ltd.

Suda, I., Oki, T., Masuda, M., Kobayashi, M., Nishiba, Y. and Furuta, S. 2003. Physiological functionality of purple-fleshed sweet potatoes containing anthocyanins and their utilization in foods. Japan Agricultural Research Quarterly 37: 167–173.

Suda, I., Ishikawa, F., Hatakeyama, M., Kudo, T., Ito, A., Yamakawa. O. and Horiuchi. 2008. Intake of purple sweet potato affects on serum hepatic biomarker levels of healthy adult men with borderline diabetes. European Journal of Clinical Nutrition 62: 60–67.

Suh, H. J., Kim, J. M. and Choi, Y. M. 2003. The incorporation of sweet potato application in the preparation of a rice beverage. International Journal of Food Science and Technology 38: 141–151.

Swain, M. R., Mishra, J. and Thatoi, H. 2013. Bioethanol production from sweet potato (*Ipomoea batatas* L.) flour using co-culture of *Trichoderma* sp. and *Saccharomyces cerevisiae* in solid-state fermentation. Brazilian Archives of Biology and Technology 56(2): 171–179.

Syamala, B. 1997. *Asparagus*—an antacid and uterine tonic. pp. 8. Science Express dt.1.7.1997.

Tan, H. Z., Gu, W. Y., Zhou, J. P., Wu, W. G. and Xie, Y. L. 2006. Comparative study on the starch noodle structure of sweet potato and mung bean. Journal of Food Science 71(8): C447–55.

Taneya, M. L., Biswas, M. M. and Ud–Din, M. S. 2014. The studies on the preparation of instant noodles from wheat flour supplementing with sweet potato flour. Journal of the Bangladesh Agricultural University 12(1): 135–42.

Teow, C. C., Truong, V., McFeeters, R. F., Thompson, R. L., Pecota, K. V. and Yencho, G. C. 2007. Antioxidant activities, phenolic and β-carotene contents of sweet potato genotypes with varying flesh colours. Food Chemistry 103: 829–838.

Terahara, N., Matsui, T., Fukui, K., Matsugano, K., Sugita, K. and Matsumoto, K. 2003. Caffeoylsophorose in a red vinegar produced through fermentation of sweet potato. Journal of Agricultural and Food Chemistry 51: 2539–2543.

Terahara, N., Konczak, I., Ono, H., Yoshimoto, M. and Yamakawa, O. 2004. Characterization of acylated anthocyanins in callus induced from storage root of purple sweet potato, *Ipomoea batatas* L. Journal of Biomedicine and Biotechnology 5: 279–286.

Teramoto, Y., Hano, T. and Udea. S. 1998. Production and characteristics of traditional alcoholic beverage made with saccharifying agent. Journal of the Institute of Brewing 104: 339–341.

Thomas, O. E. and Adegoke, O. A. 2015. Toxicity of food colours and additives: A review. African Journal of Pharmacy and Pharmacology 9 (36): 900–914.

Truong, V. D. and Avula, R. Y. 2010. Sweet potato purees and dehydrated powders for functional food ingredients. pp. 117–161. *In*: R. C. Ray and K. I. Tomlins (eds.). Sweet Potato: Post Harvest Aspects in Food, New York: Nova Science Publishers, Inc.

Truong, V. D., Hu, Z., Thompson, R. L., Yencho, G. C. and Pecota, K. V. 2012. Pressurized liquid extraction and quantification of anthocyanins in purple-fleshed sweet potato genotypes. Journal of Food Composition and Analysis 26(1-2): 96–103.

Truong, V. D., Pascua, Y. T., Reynolds, R., Thompson, R. L., Palazoglu, T. K., Mogol, B. and Gokmen, V. 2013. Processing treatments for mitigating acrylamide formation in sweet potato French fries. Journal of Agricultural and Food Chemistry 62(1): 310–316.

Tzouros, N. E. and Arvanitoyannis, I. S. 2001. Agricultural produces: Synopsis of employed quality control methods for authentication of foods and application of chemometrics for the classification of foods according to their variety of geographical origin. Critical Reviews in Food Science and Nutrition 41: 287–319.

Ubalua, A. O. 2014. Sweet potato starch as a carbon source for growth and glucoamylase production from *Aspergillus niger*. Advances in Microbiology 4(12): 788.

Ugent, D., S, Pozorski. and Pozorski, T. 1982. Archaeological potato tuber remains from the Casma Valley of Peru. Economic Botany 36: 182–192.

Versari, A., Parpinello, G. P., Scazzina, F. and Rio, D. D. 2010. Prediction of total antioxidant capacity of red wine by Fourier transform infrared spectroscopy. Food Control 21: 786–789.

Verzele, M. 1986. 100 years of hop chemistry and its relevance to brewing. Journal of Institute of Brewing 92: 32–48.

Vi, L. V., Salakkam, A. and Reungsang, A. 2017. Optimization of key factors affecting bio-hydrogen production from sweet potato starch. Energy Procedia 138: 973–978.

Vinson, J. A., Dabbagh, Y. A., Serry, M. M. and Jang, J. 1995. Plant avonoids, especially tea avonols, are powerful antioxidants using an *in vitro* oxidation model for heart disease. Journal of Agricultural and Food Chemistry 43: 2800–2802.

Walker, P. B. M. 1988. Chambers Science and Technology Dictionary. Cambridge Press, UK: Chambers.

Wang, F., Jiang, Y., Guo, W., Niu, K., Zhang, R., Hou, S., Wang, M., Yi, Y., Zhu, C., Jia, C. and Fang, X. 2016. An environmentally friendly and productive process for bioethanol production from potato waste. Biotechnology for Biofuels 9(1): 50. doi:10.1186/s13068-016-0464-7.

Wang, G. L., Yue, J., Su, D. X. and Fang, H. J. 2006. Study on the antioxidant activity of sweet potato and its inhibiting effect on growth of cancer S180. Acta Nutrimenta Sinica 28: 71–74.

Wanjuu, C., Abong, G., Mbogo, D., Heck, S., Low, J. and Muzhingi, T. 2018. The physiochemical properties and shelf-life of orange-fleshed sweet potato puree composite bread. Food Science and Nutrition 6(6): 1555–63.

Weenuttranon, J. 2018. Product development of purple sweet potato ice cream product development of purple sweet potato ice cream. International Journal of Advances in Science Engineering and Technology 6(2): 33–36.

Weirko–Manu, F. D., Ellis, W. O. and Oduro, I. 2010. Production of a non-alcoholic beverage from sweet potato (*Ipomoea batatas* L.). African Journal of Food Science 4: 180–183.

Woolfe, J. 1992. Sweet potato: An Untapped Food Resource. Cambridge, UK: Cambridge University Press.

Wrolstad, R. E. 2000. Anthocyanins. pp. 237–252. *In*: G. J. Lauro and Francis, F. J. (eds.). Natural Food Colorants. New York: Marcel Dekker.

Wu, D. M., Lu, J., Zheng, Y. L., Zhou, Z., Shan, Q. and Ma, D. F. 2008. Purple sweet potato repairs D-galactose-induced spatial learning and memory impairment by regulating the expression of synaptic proteins. Nuerobiology of Learning and Memory 90: 19–27.

Xu, S., Pegg, R. B. and Kerr, W. L. 2012. Sensory and physicochemical properties of sweet potato chips made by vacuum-belt drying. Journal of Food Process Engineering 36(3): 353–63.

Yamakawa, O. 1997. Development of new cultivation and utilization system for sweet potato towards 21st century. pp. 1–8. *In*: Proceedings of the International Workshop on Sweet Potato Production System toward the 21st Century, Miyakonojo, Miyazaki, Japan. 9–10 Dec.

Yamakawa, O. 2000. New cultivation and utilization system for sweet potato toward the 21st century. pp. 8–13. *In*: M. Nakatani and K. Komaki (eds.). Potential of Root Crops for Food and Industrial Resources. Twelfth Symposium of International Society of Tropical Root Crops (ISTRC), 10–16 Sept.

Yamunarani, S. R., Kumar, M. D., Balachander, B., Tamilarasan, K. and Muthukumaran, C. 2010. Optimization of amylase production from *Aspergillus oryzae* MTCC 1847 by submerged fermentation. Research Journal of Engineering and Technology 1(2): 79–81.

Yang, S. S. and Yuan, S. S. 1990. Oxytetracycline production by *Streptomyces rimosus* in solid state fermentation of sweet potato residue. World Journal of Microbiology and Biotechnology 6: 236–244.

Younus, S., Masud, T. and Aziz, T. 2002. Quality evaluation of market yoghurt/dahi. Journal of Pakistan Nutrition 1: 226–230.

References

Zan, Z. and Zou, X. 2013. Efficient production of polymalic acid from raw sweet potato hydrolysate with immobilized cells of *Aureobasidium pullulans* CCTC M2012223 in aerobic fibrous bed bioreactor. Journal of Chemical Technology and Biotechnology 88: 1822–1827.

Zhang, C. and Rosentrater, K. A. 2015. TEA (Techno-economic analysis) and LCA (Life cycle assessment) of small, medium, and large scale winemaking processes. 2015 ASABE Annual International Meeting. USA: American Society of Agricultural and Biological Engineers. doi:10.13031/aim.20152188570.

Index

Chapter 1

Fig. 1.3, p. 3

Chapter 7

Fig. 7.7, p. 130

a   b   c

Fig. 7.8, p. 132

Chapter 8

Fig. 8.2, p. 150

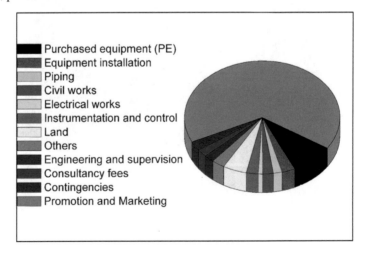

Fig. 8.3, p. 157

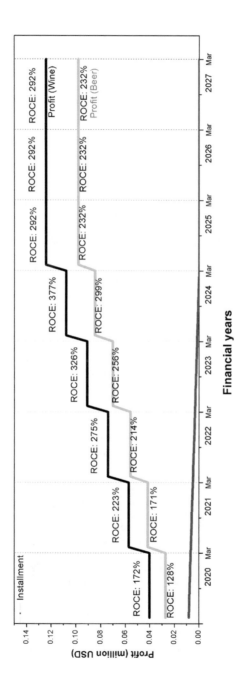

Fig. 8.4, p. 158

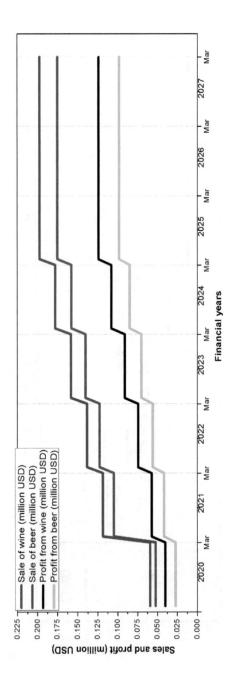